# RECENT ADVANCES IN
# ANIMAL NUTRITION — 1996

*Cover design*

James Sharp's winnowing machine, c1770

From Agricolar Sylvan Farmers' Magazine 1772

# Recent Advances in Animal Nutrition

## 1996

P.C. Garnsworthy, PhD
J. Wiseman, PhD
W. Haresign, PhD
*University of Nottingham*

NOTTINGHAM
University Press

Nottingham University Press
Manor Farm, Main Street, Thrumpton
Nottingham, NG11 0AX, United Kingdom

NOTTINGHAM

First published 1996

**British Library Cataloguing in Publication Data**
Recent Advances in Animal Nutrition — 1996:
University of Nottingham Feed Manufacturers
Conference (30th, 1996, Nottingham)
I. Garnsworthy, Philip C. II. Wiseman, J. III. Haresign, W.

ISBN 1-897676-034

Typeset by Nottingham University Press, Nottingham
Printed and bound by Redwood Books, Trowbridge, Wiltshire

# PREFACE

This book contains the proceedings of the 30th University of Nottingham Feed Manufacturers Conference. Having reached such an auspicious milestone, it is appropriate to refer back to the proceedings of the first conference held in 1967. The following is an extract from the preface of that publication.

"It was thought desirable to organise a Nutrition Conference for Feed Manufacturers along similar lines to the ones held annually at several American Universities. The conference was planned to concentrate on issues of particular interest to feed manufacturers and it was hoped to provide information of a type that could be readily applied to a commercial situation. Every effort was made to encourage uninhibited discussion and to ensure that some of the conventional restrictions of scientific communication did not impede a reasoned presentation of practical issues in the topics chosen. In particular it was hoped that opinions could be expressed without an obligatory requirement for published supporting evidence."

These aims are virtually the same today. It is a tribute to the members of the original organising committee, and those who have served on it since, that such a varied and stimulating programme can be devised each year. Further tribute must be paid to the authors who present papers and provide written manuscripts in the form you see here. Together, these people have helped make the "Nottingham Conference" famous throughout the world.

The thirtieth conference started with an update on legislation for feed compounders covering a broad spectrum of new and amended regulations. This was followed by a lively debate on the use of antibiotics in farm animals. It was acknowledged that the general public is badly informed on the use of antibiotics, particularly on their benefits in promoting animal welfare. However, the increase in resistant strains of bacteria causes serious concerns for human health and great care needs to be exercised when using antibiotics for animals.

Refinements in the feeding of poultry were discussed in papers covering amino acid profiles and phosphorus nutrition. The availability of synthetic amino acids has greatly improved our ability to balance the supply of amino acids to broilers and turkeys for optimum performance. Phosphorus is an essential nutrient for poultry but increasing requirements for pollution control mean that excesses can no longer be tolerated and phosphorus nutrition needs to be evaluated with more accurate systems, such as that presented here.

Quality control is very important for feed manufacturers at all stages of production from raw materials to finished product. Traditional methods of laboratory analysis are no longer acceptable for analysing the number of samples and range of parameters required today. Near-infrared spectroscopy has the potential to meet the requirements of speed, reliability and cost-effectiveness for modern quality assurance. In addition to its role as a replacement for traditional methods, novel applications of near-infrared spectroscopy were also reviewed.

The papers on ruminant nutrition focused on dairy cows. Firstly, milk analysis was considered as a simple, non-invasive method for determining nutritional and disease status. Secondly, the importance of grass availability for cows of high genetic merit was emphasised. Thirdly, supplementation of maize and wholecrop silages was reviewed. These three papers may not be directly related to the sale of compound feed by feed manufacturers, but they do remind us of the importance of considering the cow's response to the whole diet.

The nutrition of pigs is reaching new levels of sophistication. Energy evaluation systems are now robust enough to be able to formulate rations to meet net energy requirements. New data show that modern genotypes partition energy and protein differently, so it is necessary to re-evaluate their amino acid requirements. Further evidence suggests that we need to alter allowances when the immune system of pigs is activated by exposure to antigens.

It has been said that this collection of papers is one of the best over the thirty years of the conference. It certainly meets the criteria of being topical, interesting to feed manufacturers and readily applicable. We would like to thank the authors for their presentations and manuscripts. Thanks are also due to Trouw Nutrition for their continued financial support of the conferences.

P.C. Garnsworthy
J. Wiseman
W. Haresign

# CONTENTS

# I

# Legislation and Health

1

# LEGISLATION AND ITS EFFECT ON THE FEED COMPOUNDING INDUSTRY

B. G. VERNON[1], J. NELSON[2] AND B. C. COOKE[3]
[1] *BOCM PAULS Ltd, P. O. Box 39, 47 Key Street, Ipwsich, UK*
[2] *UKASTA, 3 Whitehall Court, London, UK*
[3] *Dalgety Agriculture Ltd, 180 Aztec West, Almondsbury, Bristol*

New legislation and amendments to current legislation are discussed under the various headings given in the text. The topics presented cover a broad selection of legislation set in place during 1995 as well as issues that are currently under discussion with the relevant legislative bodies.

## Feeding Stuffs Regulations 1995

The amendments to the Feeding Stuffs Regulations (1991), which were previously discussed (Vernon, Nelson and Ross 1995), have now been consolidated into the 1995 issue of these regulations.

### ENZYMES AND MICRO-ORGANISMS

In these new regulations the sale, supply and use of enzymes and micro-organisms (defined as additives in the Feeding Stuffs Regulations) has been legitimised. As part of the authorization procedure for enzymes and micro-organisms, manufacturers have had to submit to the authorising bodies of each Member State identification notes on the individual additives by the 31st December 1995. From the 1st January 1996 any enzymes and micro-organisms incorporated into an animal feedingstuff must be declared as per the Feeding Stuffs Regulations (1995). At the time of writing the official list of authorised products has not been published.

Where there are any doubts concerning the validity of the products then the supplier of the additive should be contacted to ascertain the position on their dossier submission. It should be noted that a Company must be able to demonstrate that they have taken steps to comply with the legislation, since no person shall sell, or have in possession with a view to sale for use as a feedingstuff, or use as a feedingstuff, any material containing an enzyme or micro-organism, or sell, or have in possession with a view to sale for incorporation in a feedingstuff, unless they are authorised.

The specific declaration requirements are listed in the Feeding Stuffs Regulations (1995) and a summary of these has been previously published (Vernon *et al.,* 1995). It is put forward as a suggestion only that, as with the declaration of vitamin E, the feed compounder declares the amount added since there is no maximum permitted level for enzymes and/or micro-organisms stated in the legislation.

## Dietetic Feedingstuffs

The consolidated Feeding Stuffs Regulations (1995) contain the requirements listed in the EU Directive on Dietetic Feedingstuffs that was adopted in July 1994. A dietetic feedingstuff is a feed which has a particular nutritional purpose. The definition of a 'particular nutritional purpose' means the purpose of satisfying the specific nutritional needs of certain pets and productive livestock whose process of assimilation, absorption and metabolism could be temporarily impaired or is temporarily or irreversibly impaired and therefore able to derive benefit from the ingestion of feedingstuffs appropriate to their condition. The declaration requirements are laid down in Schedule 1 paragraph 17/18 of the regulations. It is important to note the content of paragraph 17(2) and paragraph 18 (1/2):

17 (2) Where particular attention is drawn to the presence or low content of any ingredient as permitted by subparagraph 17(1), the minimum or maximum content, expressed in terms of the percentage by weight of the ingredient, shall be clearly indicated-

   a)    opposite the statement which draws attention to the presence or low content, or
   b)    in the list of ingredients, or
   c)    by mentioning that presence of low content and the percentage thereof (by weight) opposite the corresponding category of ingredients.

18 (1) Subject to subparagraph (2) below, in the case of any feedingstuff intended for a particular nutritional purpose the following particulars shall be contained in the statutory statement:

   a)    the term 'dietetic'

b)    a description of the feeding stuff

c)    the particular nutritional purpose of the feeding stuff, as specified in column 1 of chapter A of schedule 10

d)    the essential nutritional characteristics of the feeding stuff, as specified in column 2 of chapter A of schedule 10

e)    the declarations, prescribed in column 4 of chapter A of schedule 10

f)    the declarations, if any, prescribed in column 6 of chapter A of schedule 10

g)    where any declarations prescribed in the column do not include a declaration that it is recommended that the prior opinion of a veterinarian be sought, the words 'it is recommended that a specialist's opinion be sought before use'; and

h)    the recommended length of time for the use of the feeding stuff.

18 (2) The particulars required in the subparagraph (1) above to be contained in the statutory statement shall be declared in accordance with the requirements of paragraph 3-7 and 9 of chapter B of schedule 10.

It should be noted that the above declaration requirements are in addition to the existing statutory declaration requirements; ie oil, protein, fibre, ash, vitamins/ trace elements etc.


INSPECTORS POWERS TO ENTER PREMISES

In paragraph 22 of the Feeding Stuffs Regulations (1995) the powers of inspectors to enter premises and to inspect records are described. An inspector has the power to enter at all reasonable times any premises

a)    on which he has reasonable cause to believe any compound feeding stuff is manufactured, or

b)    which he has a reasonable cause to believe are occupied by a person engaged in the manufacture of any compound feeding stuff, for purposes related to such manufacture by him, and may on those premises -

i)    require any person engaged in the manufacture of any compound feeding stuff to produce any record, in written or any other form, relating to the manufacture by that person of any compound feeding stuff

ii)    inspect and take copies of any such record;

iii)    where any such record is kept by means of a computer, have access to any computer and any associated apparatus or material which is or has been in use in connection with the record; and

iv)    where any such record is kept as aforesaid, require any person having charge of, or otherwise concerned with the operation of, the computer, apparatus or materials to afford him such assistance as he may reasonably require

LACOTS officers have carried out visits, under both this provision and the Food Safety Act (1990), to inspect feed compounding establishments in 1995. These visits are an extension to audits already being carried out at human food establishments. The feedback from the visits has confirmed that the feed compounding process from raw material purchase to delivery of finished feed has fully complied with the Feeding Stuffs Regulations (1995) and the Food Safety Act (1990).

## IODINE IN ANIMAL FEEDINGSTUFFS

In part VI of Schedule 4 in the Feeding Stuffs Regulations (1995) the maximum iodine content in the complete feeding stuff is defined as a total of 40 mg/kg for species other than horses; the total for horses is 4 mg/kg. A proposal has been put forward to reduce the maximum level in the complete feeding stuff to:

10 mg/kg for all species
except for
20 mg/kg for fish feed
4 mg/kg for horses

In the case of ruminants, the UK guidelines for the daily iodine allowance (ADAS, 1983) is given below:

| | |
|---|---|
| winter | 0.5 mg iodine/ kg dry matter |
| summer | 0.15 mg iodine/ kg dry matter |
| presence of goitrogens | 2.0 mg iodine/ kg dry matter |

If an assumption is made that there is negligible iodine available from forage then at an average dry matter intake of 20 kg per day the new proposed maximum of 10 mg iodine/kg complete feedingstuff (basis 12% moisture) this would give an intake of 227 mg iodine on a dry matter basis. There are indications from UK farming practices that supplemental iodine is being routinely fed to ensure normal metabolism and health which exceed the proposed calculated maximum intake of

227 mg iodine per day on a dry matter basis; viz-a-viz 320 - 850 mg per day. At the current 40 mg iodine/kg complete feed (basis 12% moisture) and 20 kg per day dry matter intake the current iodine maximum intake would be 909 mg iodine on a dry matter basis.

It is important to note that over the last few years that there has been little change in the iodine level in milk ( Ministry of Agriculture Fisheries and Food (MAFF) personal communication). Thus where supplemental iodine is required to ensure normal metabolism and health in ruminants and this supplementation would exceed the proposed 10 mg iodine/kg complete feedingstuff (basis 12% moisture) there is, at the time of writing, no mechanism to allow for these increased levels of iodine to be added to say compound feed, feed blocks etc. Data are being collated and where appropriate this will be submitted to aid the discussion on the proposed reduction in iodine level for species other than for fish and horses.

## Nickel in Fats and Oils

In 1995 two Member States reported high levels of nickel in fats and oils used in animal feedingstuffs, which were stated to be connected with a decrease in animal performance. The nickel levels that were reported either by the Member States or that analysed by outside bodies ranged between 350-1300 ppm nickel. There are a number of possible sources of nickel in fats and oils:

   i)  inherent in the fats/oils
   ii) hardened fats/oils using nickel and hydrogen
   iii) tank bottoms/residue fats

The nickel level in some fats/oils used either as straights or in fat blends incorporated into animal feeding stuffs is given in Table 1.1.

**Table 1.1** THE NICKEL CONTENT OF FATS/OILS

| Fat/oil | Nickel (ppm) | Average (ppm) |
|---|---|---|
| PFAD | 0 - 6 | 1.4 |
| Fish acid oil | 0 - 6.5 | 0.01 |
| Soya oil | 0 - 2 | 0.002 |
| Tallow | 0 - 1 | 0.006 |
| RVO | 0 - 5 | 0.13 |
| Acid oils | 0 - 1 | 0.005 |
| Mixed soft | 0 - 0.5 | 0.002 |
| Residue (not hardened) | 50 - 80 | |
| (hardened) | > 2000 | |

Analysis of fat blends incorporated into animal feedingstuffs in the UK indicate levels of less than 2 ppm nickel. It should be noted that many of the fats/oils (other than the residue material) listed in Table 1.1 are also consumed by humans; ie cooking oil, margarine, foods. In the case of the residue material, research within the UK in the early 1980s indicated that the inclusion of such material into animal feedingstuffs had a negative effect on nutrient availability (Vernon and Cooke unpublished data; Edmunds 1989).

The data gleaned in the UK will be forwarded to the EU Commission who are reviewing the issue of nickel in animal feedingstuffs.

## Directive for fixing the principles governing the organisation of official inspections in the field of animal nutrition 1995

This Directive was agreed at the Council of Ministers in October 1995 and all Member States must implement this by April 1998. This Directive is designed to fix harmonised principles governing the organisation of inspections in the field of animal nutrition. This would cover all the stages from the production, processing, storage, transport, distribution trade and use of feedingstuffs, as well as imports into the EU from Third World Countries. Each Member State must produce an annual report by not later than April 2000 which will list as a minimum:
i)      the number and type of inspections carried out
ii)     the results of the inspections, in particular the number and type of infringements found
iii)    actions taken where infringements have been found

The check function will cover existing legislation such as the Directives on additives, maximum permitted levels for undesirable substances and products in feedingstuffs, marketing of animal feedingstuffs, certain products in animal nutrition, feedingstuffs intended for a particular nutritional purpose (dietetic feeds) and future legislation. This check function could also apply to other legislation such as the UK codes of practice on the control of salmonella. The checks referred to above will cover such aspects as:

i)      documentation to verify the product
ii)     identification checks to confirm the product documentation, including labelling
iii)    physical check to include analytical testing

Where checks show that the products do not meet the requirements, Member States shall complete a number of procedures which will apply to either trade imports from Third Countries or trade within the Community. In the case of imports the Member States may prohibit their entry and marketing as well as redispatch out of the Community any such material. Alternatively, the competent authority can authorise, in the case of imports and materials traded within the Community, one of the following actions:

i)  bringing the products into line with the requirements within a deadline to be determined; or
ii)  decontamination where appropriate; or
iii)  processing in any other suitable manner; or
iv)  use for other purposes; or
v)  destruction of the products.

Where a Member State identifies a problem other Member States must be quickly notified of the nature and outcome of the checks that have been carried out, the decisions taken and the reasons for such decisions.

## Directive laying down the conditions and arrangements for approving and registering certain establishments in the animal feed sector and other EC business 1995

This Directive was agreed at the Council of Ministers in September 1995 and all Member States must implement this by April 1998. The registration refers to establishments manufacturing additives, premixtures and compound feedingstuffs containing premixtures or additives. In addition, it could cover home-mixers who circulate feed to third parties or use premixtures or additives. Intermediaries in the distribution chain would also be required to register. Also any Company that operates on several sites will have to register each establishment individually. It should be noted that the term additive includes certain elements and substances listed in the Feeding Stuffs Regulations (1995), i.e. copper, selenium, enzymes and micro-organisms plus those authorised for incorporation into animal feedingstuffs under the Medicine Act (1968) and all its amendments.

As part of this Directive there are new measures required for, respectively, European manufacturers of certain non-medicinal additives as well as manufacturers of medicinal and non-medicinal additives based in Third Countries. Such a Directive should not result in an imbalance of trade both within and into the EU. This is why the MAFF have indicated that there will be an on-going discussion on the interpretation of this new Directive in conjunction with existing Feeding Stuffs Regulations and the Medicine Act requirements.

## Bovine Spongiform Encephalopathy (BSE)

Throughout 1995 there has been continuing investigation into the origin of BSE as well as cooperation between MAFF, UKASTA, feed compounders and the rendering industry on the various new and exiting legislation related to BSE. In this section a number of pieces of legislation are discussed.

### SPECIFIED BOVINE OFFAL ORDER 1995

This Order consolidates and streamlines the existing rules on Specified Bovine Offal (SBO). The main items of the Order are as follows:

i)      to require all rendering plants, head-boning plants, incinerators and collection points for SBO to be approved by Ministers. Rendering plants will have to have a totally dedicated processing line for SBO;

ii)     movement permits for SBO issued by local authorities will no longer be required in view of the tighter controls being introduced on approval and record keeping throughout the disposal chain;

iii)    the brains and eyes of a bovine animal of six months or more may no longer be removed from the head so that the part of the skull containing the brain and eyes, after removal of the head meat and tongue for human consumption, will have to be disposed of as SBO.

The requirement for rendering plants to have a dedicated line must be in place and approved by MAFF by February 1996. This requirement for processing SBO should also be viewed in the light of the legislation put in place for the heat treatment systems used in the processing of animal waste of ruminant origin (EC, 1994). This latter decision lays down time/temperature profiles and other parameters for the rendering of ruminant meat and bone to ensure the inactivation of spongiform encephalopathy agents. This was reviewed and summarised in a previous presentation (Vernon *et al.*, 1995).

### COMMISSION DECISION 95/287 ON BSE

This Commission Decision on BSE requires the UK to introduce a formal monitoring programme to test for the inclusion of ruminant protein in ruminant feedingstuffs. The Decision was based on several pieces of evidence, some incomplete, that indicate the transmission of infected tissue via feed as a potential cause of BSE cases born after the ruminant feed ban (BAB). It requires the targeting of not only feed compound mills that produce monogastric feeds but all those manufacturing ruminant diets. It should be noted that there is also the continuing on-farm BSE and BAB investigations that involve an audit of the straights, blends and compound feedingstuffs that have been fed to ruminant animals.

The sampling and monitoring programme will incorporate the ELISA technique developed by MAFF to detect the presence of ruminant proteins in rendered animal material (Ansfield, 1994). The technique can differentiate between ovine/bovine and porcine meat and bone in animal feedingstuffs.

Prior to implementing the Decision a UK-wide mill survey has been carried out. The collated information will be used to prepare a sampling frequency based on a total tonnage turnover. From this evaluation a mill could be visited once per week through to once per month; there will be a minimum flat rate monthly sampling protocol. Samples of finished feed will be taken in a prescribed manner, split and clearly identified in sealed containers for analysis at Luddington. At the time of writing MAFF have indicated that this formal sampling procedure will commence at the end of January 1996. Procedures have been put in place for reporting the results and actions, if necessary, in the event of a positive ELISA test have been laid down.

MAFF have reported that the BSE epidemic is clearly in decline. As predicted, fewer younger animals are becoming infected and this, taken together with the general decline in the number of cases in all age groups of cattle, confirms the success that the combined approach of MAFF, the compound trade as a whole, the rendering industry and the farming community are having in reducing the incidence of BSE and BAB animals.

## Establishment of Maximum Residue Limits (MRLs) under Council Regulation 2377/90

Council Regulation 2377/90 provides an EU framework for the setting of MRLs for residues of veterinary medicinal products incorporated into animal feedingstuffs. Under this regulation MRLs will be established for all the pharmacologically active substances used in food producing animals by the end of 1996. After this time, Member States will not be able to issue a marketing authorization (a product licence in the UK) for use in such an animal and will have to revoke existing licences, unless the substance is listed in one of the following Annexes to the Regulation:

Annex I:    substances for which a full MRL has been fixed;

Annex II:    substances for which an MRL is not required;

Annex III:    substances for which a provisional MRL has been fixed;

Annex IV:    contains those substances for which no MRL can be fixed on consumer safety grounds and therefore effectively prohibited from food producing animals.

It should be noted that any existing UK marketing authorizations for products containing substances in Annex IV would be revoked under the Medicine Act (1968).

It is important to understand that MRLs are assessed on toxicological and residue grounds (ie safety, efficacy and environmental) to safeguard human health and are scientifically based on diet tolerance assumptions. There are considerable safety margins applied when setting MRLs. These MRLs form part of the Member States National Surveillance Survey approach to food safety. The establishment of MRLs should be used as an aid in the interpretation of such data to overcome public perceptions of risks associated with food. To date MRLs for some 180 products have defined and agreed.

## Code of Practice for the Control of Salmonella 1995

The revised codes of practise for the control of salmonella were adopted in June 1995 and have been published. These cover the control of salmonella

i)      during the storage, handling and transport of raw materials intended for the incorporation into, or direct use as, animal feedingstuffs
ii)     in the production of the final feed for livestock in premises producing less than 10,000 tonnes per annum
iii)    in the production of the final feed for livestock in premises producing over 10,000 tonnes per annum

In essence they are similar to the previous codes of practice but there is now the emphasis on the use of the principle of Hazard Analysis Critical Control Point (HACCP) investigations. The HACCP principles described in the codes of practice are as follows:

i)      all areas and process are assessed for 'risk';
ii)     regular surveillance of certain Critical Control Points(CCP) are carried out, with particular reference to non-compliance at each CCP. Each non-compliance should be recorded and acted upon using a non-compliance procedure assigned to each CCP;

iii)   records of all HACCP systems should be kept for observation and inspection by the authorities.

As part of the new codes of practice the reporting of salmonella positives in finished feedingstuffs has been changed. Previously under the heading of pig compounds and poultry compounds the total positive results by species for both meal and pelleted products were reported. As from June 1995 the meal and pellet products have been split. The categories that are now published are pig/poultry meals, pig extrusions (pelleted) and poultry extrusions (pelleted). A comparison of the salmonella positives in finished feed over the last three years are presented in Table 1.2.

**Table 1.2** SALMONELLA POSITIVES (%) IN FINISHED PRODUCTS

|                      | *1993* | *1994* | *1995* |    |
| -------------------- | ------ | ------ | ------ | -- |
| Ruminant             | 4.7    | 3.4    | 2.8    |    |
| Protein concentrates | 3.2    | 3.7    | 7.4    |    |
| Minerals/others      | 0.4    | 0.5    | 0.4    |    |
| Pig compound         | 2.1    | 3.6    | 3.9    | *  |
| Poultry compound     | 2.7    | 2.7    | 3.4    | *  |
| Pig/poultry meal     |        |        | 2.8    | $  |
| Pig extrusions       |        |        | 0.9    | $  |
| Poultry extrusions   |        |        | 0.6    | $  |

\*   January - June 1995
$   June - December 1995

Source: MAFF 1993; MAFF 1994a; MAFF 1995

The new reporting procedure shows up a difference between meal and pelleted feeds. Between January-June 1995 there appeared to be a similar level of salmonella positives in pig compounds and poultry compounds. The data for the period June-December 1995 which splits the meal and pelleted feeds shows a different pattern. The low level of positives in poultry extrusions (0.6%) is likely to reflect the higher processing temperatures compared to pig extrusions (0.9%). The significantly higher positives in pig/poultry meals (2.8%) may be a reflection of the need to carry out an HACCP analysis at a mill(s). A similar conclusion could be drawn from the protein concentrate results since the majority of these products are manufactured as meals.

Investigations by MAFF and the feed compound industry as a whole into the source and serotypes of salmonella found in raw materials, finished products and animals forms part of the continuing approach to food safety surveillance. A recent study by MAFF indicated that whilst salmonella was not ubiquitous when it was present on-farm it was widespread, as indicated in Table 1.3.

**Table 1.3** SALMONELLA SURVEY

|  | Cattle environment | Grain trailer | Grain storage | Manure (field) | Wild birds |
|---|---|---|---|---|---|
| Farm |  |  |  |  |  |
| A | Pos | Neg | Pos | Pos | Pos |
| B | Neg | Neg | Pos | Neg | Neg |
| C | Pos | Neg | Pos | Pos | Pos |
| D | Pos | Neg | Neg | Pos | Neg |
| E | Pos | Pos | Pos | Pos | Pos |

Pos = positive
Neg = negative

Whilst there are many hundreds of salmonella serotypes which are reported as part of the salmonella codes of practice MAFF take particular note when *s. typhimurium, s. derby* and *s. enteritidis* are found in raw materials and animal feedingstuffs. It is interesting to note the top four serotypes found in raw materials, pig feed per se and pigs/pig meat (Table 1.4, MAFF 1994b)

**Table 1.4** SALMONELLA SEROTYPES FOUND IN RAW MATERIALS, FEED AND PIG MEAT

| Raw Materials | Pig Feed | Pigs/Pig Meat |
|---|---|---|
| s. tennessee | s. seftenberg | s. typhimurium |
| s. montivedeo | s. tennessee | s. derby |
| s. mbandaka | s. melagridis | s. panama |
| s. seftenberg | s. binza | s. kedougou |

Source: MAFF (1994b)

In all three cases the top four serotypes accounted for 65%, 74% and **88%** of the serotypes found in raw materials, pig feed and pigs/pig meat respectively. It is interesting to note that both *s. tennesse* and *s. seftenberg* were high in raw materials and they have come through into the finished feed. The difference is the serotypes found in the pigs/pig meat where the predominant serotype was *s. typhimurium*. The level of *s. typhimurium* as a percentage of the total number of positives reported by MAFF (MAFF 1994a; MAFF, 1995) in raw materials and finished products was 1.1% and 2.5 % respectively in 1994 and 1.9% and 2.3% respectively in 1995. The *s. typhimurium* in pigs/pig meat referred to above accounted for approximately two-thirds of the positives.

Thus when consideration is given to the sources and serotypes of salmonella found in animals and human food then all origins of the contamination should be investigated from raw materials through to the preparation and packaging of human food.

## Draft control of pathogens in raw materials of vegetable origin and animal feedingstuffs

This is a draft Commission proposal on the control of salmonella in raw materials and finished products. It is based upon the current UK codes of practice but are for implementation throughout all Member States.

The level of salmonella positives in raw materials over the last three years are presented in Table 1.5.

**Table 1.5** SALMONELLA POSITIVES (%) IN RAW MATERIALS

|  | *1993* | *1994* | *1995* |
|---|---|---|---|
| Animal protein (UK processing plants) | 2.2 | 2.2 | 1.9 |
| GB/imported animal protein | 2.7 | 4.1 | 4.4 |
| Linseed/rape/soya/sun (UK origin) | 3.9 | 4.7 | 2.7 |
| Other oilseeds/products | 7.9 | 4.9 | 3.5 |
| Non oilseeds | 2.1 | 2.0 | 1.5 |

Source: MAFF 1993; MAFF 1994a; MAFF 1995

The level of salmonella positives in animal proteins (UK origin) and non-oilseed raw materials like cereals, wheatfeed, rice bran and sugar beet pulp continue to have the lowest risk factor. The level of positives in the GB/imported animal proteins should be viewed in the light of the level of positives reported for imported fishmeal (which are made available to the market) which is reported to be less than this value (I. Pike (IFOMA) personal communication). The level of positives in oilseed products does give cause for concern. This proposed EU legislation will aid in the aim of reducing the level of salmonella in these type of raw materials.

## Workplace Regulations 1992

Under these Regulations, in particular Regulation 13, where a person might fall greater than 2 metres then there is a requirement to provide secure fencing. This

Regulation comes into operation on 1st January 1996. In relation to the sheeting of lorries, however, a proposal has been put forward that the installation of easy sheets on both lorries and trailer units for the delivery of raw materials to a mill and finished products on-farm should be in place no later than August 1996. The installation of easy sheets should apply not only to each individual Company's vehicles but also to third party hauliers.

## Dust explosions

There is currently on-going discussion with the Health and Safety Executive (HSE) on the production of a guidance note to cover both existing and new legislation on the control of dust explosions in feeds mills. The guidance note will form the basis for inspection by the HSE on the design, modification and installation of new equipment in compound feed mills.

## Acknowledgements:

The supply of information and advice on the nickel and iodine issues from Mr S. Lowe of FF Man Feed Products and Dr J. D. Allen of F Wright Ltd, respectively, are gratefully acknowledged.

## References

ADAS (1983) Mineral, Trace Element and Vitamin Allowances for Ruminant Livestock. London: Her Majesty's Stationary Office

Ansfield. M (1994) Production of a sensitive immunoassy for detection of ruminant proteins in rendered animal material heated to >130°C. Food & Agricultural Immunology, 6, 419-443

Code of Practice for the Control of Salmonella 1995. London: Her Majesty's Stationary Office

Commission Decision 95/287 on BSE (1995). EC Directive

Directive for Fixing the Principles Governing the Organisation of Official Inspections in the Field of Animal Nutrition 1995. EC Directive

Directive Laying Down the Conditions and Arrangements for Approving and Registering Certain Establishments in the Animal Feed Sector and Other EC Business 1995. EC Directive

EC (1994) Approval of Alternative treatment systems for processing animal waste of ruminant origin with a view to the inactivation of spongiform encephalopathy agents (Commission decision) 1994. EC Directive

Edmunds. B. K. (1989) Chemical Analysis of Lipid Fractions. Feedstuffs Evaluation. 50th Easter School. p 197-213. Eds Wiseman. J and Cole D. J. A.. Publishers Butterworths

Establishment of Maximum Residue Limits (MRLs) under Council Regulation 2377/90 (1990). EC Regulation

The Feeding Stuffs Regulations (1991). London: Her Majesty's Stationary Office

The Feeding Stuffs Regulations (1995). London: Her Majesty's Stationary Office

Food Safety Act (1990). London: Her Majesty's Stationary Office

MAFF (1993) Salmonella in Animal Feedingstuffs and Ingredients 1993 (Jan-Dec). London: Her Majesty's Stationary Office

MAFF (1994a) Salmonella in Animal Feedingstuffs and Ingredients 1994 (Jan-Dec). London: Her Majesty's Stationary Office

MAFF (1994b) Salmonella in Animal and Poultry Production 1994. London: Her Majesty's Stationary Office

MAFF (1995) Salmonella in Animal Feedingstuffs and Ingredients 1995 (Jan-Dec). London: Her Majesty's Stationary Office

The Medicines Act (1968). London: Her Majesty's Stationary Office

Specified Bovine Offal Order (1995). London: Her Majesty's Stationary Office

Vernon. B. G., Nelson. J. and Ross. E. J. (1995) Legislation and its effect on the feed compounder. Recent Advances in Animal Nutrition 1995. Eds Garnsworthy. P. C. and Cole. D. J. A.. Publisher: Nottingham University Press

Workplace Regulations (1992). London: Her Majesty's Stationary Office

# 2

# BENEFITS OF ANTIBIOTICS IN ANIMAL FEED

J. R. WALTON
*The University of Liverpool, Veterinary Teaching Hospital, Leahurst, South Wirral, L64 7TE, UK*

## Introduction

All species of animals are at risk from infection by micro-organisms at any stage in their life span. There are obvious high risk points from birth to slaughterweight when certain high stress events occur and it will be at these times that individuals and herds will be at greatest risk from becoming infected. Such high stress events for the pig include the immediate neo-natal period, weaning, dietary change, any relocation to a new environment, vaccination, tooth clipping and tail docking. With poultry, the high stress events will also include the immediate post-hatching period, vaccination, dietary change and relocation. Management tries as far as possible to minimise the level of stress at these high risk times and in this is generally very successful. Unfortunately, the non-farming public and often the non-specialist press retain and broadcast the erroneous opinion that all pigs and poultry are reared from birth on feed containing antibiotics which are also used for the treatment of disease in man and animals. It is imperative that there is a clear understanding of the types of antibiotics (antimicrobials) that are used in animal feed and the relationship between these and antibiotics that are used for the treatment and prevention of disease in human and animal medicine. As important is an understanding of why antibiotics are given to food producing animals, especially when it does not involve the treatment and prevention of disease.

The Swann Report in 1968 defined two classes of antibiotic for food animal use (1) Therapeutic and (2) Feed. The criteria used for this division derive especially from the need to protect human safety, and thus residues and antibiotic cross resistance are major components in the classification. Obviously a true realisation of the nature of antibiotics, how they work and why they are used in animal agriculture is paramount to the understanding of their use in food producing animals.

## WHAT IS AN ANTIBIOTIC?

Antibiotics are substances that can deleteriously affect bacteria either by interfering with their growth, metabolism or actually killing them in situ. This presentation will use the words antibiotic and antimicrobial synonymously and will avoid totally the use of chemotherapeutic. Some antibiotics are man-made but the majority are isolated from bacteria or fungi that grow naturally in our environment so it is highly likely that many bacteria have already been in contact with one or more antibiotics that are currently being used for therapeutic purposes in human and veterinary medicine. This has resulted in confusing attempts to differentiate between an antibiotic, an antimicrobial and sometimes, a chemotherapeutic. Several antibiotics are manipulated during their manufacture to add on chemical elements to protect the antibiotic against destructive bacterial enzymes.

## WHY DO WE USE ANTIBIOTICS?

Antibiotics are used in circumstances when the presence or overgrowth of specific bacteria would be detrimental to animal health. Antibiotics are also used to manipulate bacteria that are causing a reduction of the physiological or metabolic performance of food producing animals. Antibiotics are added to the feed of animals because, for convenience, a very large number of them can be medicated at the same time, it avoids the stress of handling that would be required for individual dosing and in the herd situation, it is most important to get all the animals at risk treated in an identical manner with the same product at the same time. Whilst companion and competition animals are also given antibiotics in the feed, the technique is primarily used for food producing animals, particularly pigs and poultry, that are usually housed indoors in large numbers.

## WHERE DO ANTIBIOTICS COME FROM?

Some antibiotics are man-made but the majority are isolated from bacteria or fungi that grow naturally in our environment. This latter has specific implications for the origins of resistance to antibiotics in both human and animal bacteria.

## WHY USE ANTIBIOTICS IN LIVESTOCK FARMING?

Antibiotics are used in livestock farming for many reasons other than for the treatment of disease. This is necessary because, unlike human medicine, not only has the value of the animal to be taken into consideration but also the extent and efficiency of its performance - be this feed conversion, daily live weight gain or laying eggs - all of which can be measured and improved using performance enhancers such as feed antibiotics, enzymes and new feed presentation methodology. Research is constantly yielding innovative technologies which must be applied if the UK agricultural industry is to remain competitive in an unfortunately decreasing market. So, unlike human medicine where antibiotics are used almost entirely for the prevention and treatment of disease, in animals there are at least seven indications for antibiotic use. The most controversial and least understood, yet by far the most cost effective, is the indication for performance enhancement.

Antibiotics are used in food animals for the following reasons:

(1)   To treat disease and save lives
(2)   To prevent the development of disease
(3)   To lessen pain
(4)   To avoid secondary bacterial infection
(5)   To prevent the development of an epidemic
(6)   To stabilise the gut flora and so enable the animal to cope with regular changes in dietary ingredients
(7)   To enhance physiological and metabolic performance

   Livestock agriculture has utilised the full potential of antimicrobial substances in the above manner and in so doing has  provided numerous benefits both for man and for livestock farming.

The benefits for man include:

(1)   Cheaper food
(2)   Food that is more suited to the needs of the consumer i.e. less fat
(3)   Pigs and chickens that are reared free of chronic diseases
(4) less carcass meat is condemned because of low grade disease
(5) quicker growth rate therefore meat will be more tender

The benefits for the animals include:

(1)   Rapid treatment of bacterial disease
(2)   Reduction of low grade disease
(3)   Cessation of pain
(4)   Ability to cope with constantly changing dietary ingredients without  the development of intestinal problems

(5)     Reduction of chronic toxicity by preventing the overgrowth of intestinal
        bacteria that produce a variety of toxic substances such as ammonia and
        monoamines

The use of antibiotics in livestock agriculture has been heavily regulated both
in the EC and UK in an attempt to avoid as far as possible any problem for human
health. All antibiotics proposed for animal use have to satisfy the legislators on a
range of factors related to QUALITY, SAFETY AND EFFICACY. If these
parameters are deemed satisfactory then a product licence is granted to enable the
medicine to be used in food producing animals. Also, as the text of this paper will
show, the depth of scientific investigation for an animal-use product is far greater
than for most human-use preparations because of the need to satisfy the regulators
that human health and the environment will not be compromised. Regulation does
not stop at the licensing stage, all agricultural users of antibiotics have to comply
with a variety of legislation in respect of antibiotic administration and on-farm
records and with specified directions to feed manufacturers being issued by
veterinary surgeons for antibiotic inclusion in animal feed. Manufacturers of
medicated feed are checked on a regular basis for compliance with the appropriate
regulations. At the food production site, be it slaughterhouse or milking parlour,
samples of meat and milk are routinely tested for antibiotic residues and violations
attract a heavy financial penalty. Epidemiological investigations of violator
premises involve not only the owner of the animals or milk but include the
veterinary supplier of the antibiotics. The owner is obliged under the Fresh Meat
Regulations (1995) to record every administration of a medicine to a food animal
listing the supplier of the medicine, its batch number, amount administered etc.
etc. These records are inspected by any visiting veterinary officer of MAFF or
member of the Royal British Pharmaceutical Society to ensure the record is being
kept and that identification of treated animals and withdrawal times are being
properly managed. Finally, matters related to the in-field use of antimicrobials
and safety aspects for the human meat consumer are always under constant scrutiny.
Any potential safety matter that is identified during use of a product either through
the notification of suspect adverse reactions, by publication in the scientific press
or by interest from another European State will be considered by the Medicines
Directorate and appropriate action will be taken if required.

## ANTIBIOTICS AND SAFE FOOD PRODUCTION FOR HUMANS

Many bacteria and fungi produce a wide range of substances that  inhibit the
growth of, or even destroy, a variety of other micro-organisms. These so-called
antimicrobial substances are commonly known as antibiotics and they have been

specially adapted for use in human and veterinary medicine for the treatment and prevention of bacterial diseases. Antibiotics are also used for a variety of purposes in research, commerce and industry. In nature, bacteria and fungi produce antimicrobial substances in order to give them a competitive advantage in locations like the soil, but also in the intestinal tract and on the skin of man and animals. These antimicrobial substances are produced by bacteria during both normal metabolism and other physiological activities. For instance, when an environment becomes dry and hostile, several types of bacteria will develop a resistant phase called a spore. A penicillin-like substance can be produced by the bacterium that will cause it to split open and release the spore into the environment. The spore will now remain dormant until conditions improve when it will germinate and once again produce a viable bacterium. In this way, the continuity of the bacterial species will be ensured. Commercial involvement in the development and use of antimicrobials for disease treatment purposes was originally directed at purification and large scale production. Currently, some antimicrobials are produced artificially but strain selection and genetic engineering has enabled the yield of these antimicrobials to be increased enormously. Initially, there was serious concern that the widespread use of antimicrobials, especially in agriculture, would cause problems for their usefulness in treating human diseases. However, after several decades of antibiotic use, there is now a much better understanding about their safe use and also the previous concern has been very much reduced as a result of this experience, so that the whole area of antibiotic resistance is now no longer cloaked in misunderstanding and secrecy.

The overall use of antimicrobials in agriculture has increased yet there is no evidence of any significant long term increase in antibiotic resistance in bacteria obtained from food producing animals. The application of specific regulations has resulted in a shift in the way that antibiotics are used in certain circumstances and this has no doubt been aided by the introduction of penalties for the presence of antibiotic residues in human foods. Some farmers who use organic methods prohibit the use of antibiotics except as a last resort in certain cases of animal disease. Unfortunately, this can lead to more deaths and also to the persistence of chronic diseases with associated lowered individual animal production. Some supermarkets closely regulate antibiotic use by their food producing suppliers by insisting that animals are reared in the absence of continuous feed medication. Internationally, there have been many new standards established for the safety of antibiotics, for instance the establishment of maximum residue levels (MRL's) that are permissible in animal tissues. Numerous well established antibiotic products have been re-evaluated, some have been withdrawn because of perceived potential risks for human health and the overall use of antibiotics in animal feeds is now closely monitored by a system of producer certification. On-the-spot quality control of the whole process of medicated feed manufacture is carried out by an officially appointed national organisation namely The Royal Pharmaceutical Society of Great Britain (RPSGB). The biggest change of all however has come from consumer concern about a variety of environmental issues. These matters include pollution of waterways, application of nitrogen-based fertilizers, pesticide sprays, the production of all kinds of foods free from chemical residues and especially the methods that are used for rearing and growing animals that will ultimately be used for human food production. These environmental and consumer issues are now very high on the political agenda of many Governments and the legal use of antimicrobials and hormones in intensively-housed food animals is coming increasingly into question by welfare and political organisations alike.

This paper will (1) update a previous article (Walton,1983), (2) detail the changes that have occurred over the past decade in the use of antimicrobials in agriculture; (3) set out the extent of testing that has to be carried out to provide evidence to the regulatory authorities on the QUALITY, SAFETY AND EFFICACY of antimicrobial substances; (4) review the postulated hazards for man and discuss published evidence for and against any health risks; (5) review the legislation relating to the proper use of antimicrobials in food-producing animals and the measures applied by the Regulatory Authorities to ensure compliance; (6) detail the issues concerning tissue residues relative to the use of antimicrobials in animal production and (7) discuss how the application of animal health schemes and newer animal production systems prevent inadvertent antimicrobial residues in foods.


CHANGING METHODS IN ANIMAL FARMING

Over the past two decades, there have been significant changes in livestock farming methods brought about by scientific and technological advances, together with specific demands from the consumer of meat and meat products. At the same time, there has been an increase in the size of individual animal production units but with no general increase in the total populations of animals being produced. These changes have inevitably caused a decrease in the labour force in an attempt to keep costs down to a minimum, but most of all there appears to have been a real reduction in the number of career farmers and stockmen.This will have serious implications for the overall management of all sectors of the livestock industry. Political measures have decreased the impact of national agricultural training agencies and there has been a general disenchantment with livestock production which requires long working hours often in unsocial environmental conditions. This latter has resulted in workers that have had no previous experience of farm livestock being drawn into the industry and this can only be detrimental to the well-being of animals in general. Loss of import controls, abandonment of border checks and cessation of quarantine regulations in order to satisfy single market requirements, have all resulted in an increased health risk for UK livestock. Obviously, this will demand a greater awareness by farm management of new disease problems and a requirement for a more vigorous application of in-house disease control measures. The concern over the relaxation of animal movement restrictions has recently been realised with the introduction into the UK of Blue Ear disease in pigs, Brucellosis in milking cows, Equine Viral Arteritis and, contrary to EC policy, the importation of cattle into UK from Eastern Europe with falsified certificates in respect of Foot and Mouth vaccination. All these conditions would probably have been prevented by previously accepted restrictions on the movements of live animals into UK

Human foods are obtained from a wide range of animal species including cattle, pigs, sheep, poultry, fish and also crustaceans. All these species are intensively farmed in order to optimise productivity and to enable the individual animal to perform to its full genetic potential. There have been significant advances in the selection of specific animal strains which are highly prolific and very efficient at converting cereals and other ingredients into lean edible meat. At the same time, there have been major developments in the selection of genetic hybrids that have a lean body conformation associated with increased muscle mass. Such developments have been produced in response to consumer demand and have generally followed the development of new technologies. Regardless of the genetic superiority of these new strains they are still susceptible to infection with disease-causing bacteria and similarly the micro-organisms in their intestinal tract can become destabilised by stressors, inappropriate feed ingredients and bad husbandry methods. So despite all the advances in animal breeding and production, the need is as great as ever for effective antibiotics to treat and prevent disease in all species of animals destined for human food.

## Use of antimicrobials on farms

Farm animals are just as likely as humans to succumb to generalised infectious diseases or indeed to develop local bacterial infections. Often the most suitable method of treating this type of health problem is to medicate with an appropriate antimicrobial agent. When dealing with human bacterial infections it is often the case that only one patient is involved whereas farm animals are frequently housed together in very large numbers in one building. Any infection in one animal will often rapidly spread to many others. If health problems are recognised at a very early stage and the appropriate treatment given the animals will in most cases fully recover. One of the tasks for the farmer and his veterinary surgeon, once therapeutic medication has ceased, is to ensure that there are no significant residues of antimicrobial substances remaining in the meat or other animal produce such as milk or eggs. Initially, this may seem a relatively simple matter but since so many issues are involved and a whole range of criteria have to be satisfied, then the best means of ensuring insignificant antibiotic residues in edible tissues is probably by legislative means together with routine surveillance at all stages of the animal production cycle. As a last resort, appropriate financial penalties can be used.

There is a need to recognise that treatment of bacterial diseases requires therapeutic concentrations of antibiotics whereas there may be less need for such large amounts of antibiotics with strategic preventive medication programmes. In most animal production systems it is possible to identify situations such as relocation or dietary change which together with vaccination programmes act as stressors and initiate the development of bacterial multiplication with subsequent appearance of disease. Management policy will monitor the timing of these situations and initiate a strategic medication programme to coincide with the stress related situations. With such a programme being commenced before widespread multiplication of bacteria has occurred, it has been consistently shown that the development of certain bacterial diseases can be prevented.


ANIMAL HEALTH

Bacterial infections in animals lead to such conditions as pneumonia, pleurisy, various kinds of enteritis, meningitis, skin infections and arthritis. It is essential that these problems are treated promptly and effectively so that the sick animal will completely recover and return to full production as soon as possible. Bacterial infections must be treated with specific antimicrobial agents given either by mouth or by injection. With animals that are ill and unwilling to eat it is usual to treat with antibiotics added to the drinking water. Animals that are less affected can be treated speedily with antibiotics that are mixed with the food. In the case of fish, antibiotics can be included in the food or added to the water in which they swim and in extreme conditions fish can be individually injected in the same way as farm animals. In many situations, food animals are kept in very large groups. These intensively farmed animals are no more susceptible to bacterial infections than are free range animals. However, since farm animals are kept together in large groups then any infections, if not successfully brought under immediate control, can spread very rapidly around the herd or flock. Also, because many animals are housed together in the same air space, if one animal should show signs of disease then it is quite likely that many others will have been infected at the same time. In order to deal with all these problems successfully, antibiotic treatment must be administered promptly often to large groups of animals for several days until the infection clears and the animals return to full health and production. It must also be recognised that many bacterial infections may result in severe pain for the sick animals so rapid and effective treatment is essential to reduce this pain and improve the welfare status of the animals concerned.

### Use of in-feed and water medication

The two main methods of supplying medication to groups of farm livestock are (1) dissolving it in the drinking water and (2) including it in the feed. It is impracticable to administer medicines by injection to many thousands of animals in order to treat or prevent a bacterial disease. Penning and handling animals that are sick and have never been previously touched by humans will cause undue stress and may even result in several of the animals dying. For this reason, antimicrobial and other treatments are usually administered by mouth and not by injection. Obviously, herd medication is a skilled operation if one is to ensure that all animals obtain the requisite amount of medication for their body weight and that appropriate legislative controls are applied in order that human health is not compromised.

(a)    Medicated feedingstuffs: Antibiotics can either be added to foodstuffs at the mill or on the farm. In both cases, there is strict statutory control of the whole operation and there are severe penalties for non-compliance. Antibiotics can also be spread on the top of concentrated feed once it has been put into the food trough (this process is known as top dressing) and finally individual animals can be dosed by mouth using suitable oral dosers containing the appropriate medication. Once added to a concentrated feed, some antibiotics will become unstable and rapidly lose potency. Various pharmaceutical manufacturers have countered this problem by pre-coating the antibiotic with carbohydrate or starch that will protect the antibiotic until the protective layer becomes broken down in the gastro-intestinal tract during the process of digestion after which the antibiotic will be released. There can be a very serious drawback to using medicated feedingstuffs. Many sick animals will refuse to eat and therefore will not obtain any medication at all. Others will only eat a part of their daily allocation because of a poor appetite and so they will not get sufficient medication based upon body weight. Finally, animals that are being bullied will also be prevented from eating enough food and similarly will be deprived of adequate medication on a daily or weight basis.

(b)    Medicated drinking water: There are many situations where sick animals will refuse to eat solid food. This is an expected clinical finding when the animal has a high temperature, has lost its appetite and is disinclined to walk about. This situation is commonly seen with such diseases as pneumonia or acute enteritis. However, the animal will still continue to drink because the illness will have caused some dehydration and created a need to take in water. Medication by drinking water is probably the best method of ensuring that antimicrobials are consumed by sick animals in adequate amounts on a daily basis. Management must ensure that a water soluble antibiotic preparation is properly dissolved in the drinking water which is then made available ad libitum to all animals requiring it. The requisite amount of medication will be taken in by the animal because the drinking water will have had added to it the appropriate antibiotic at a concentration consistent with the amount of water calculated to be consumed by the individual animal on a daily basis. When prescribing a particular medication other than via medicated feeding stuffs, e.g. by inclusion in the drinking water, a veterinary surgeon must adhere to the regulations made under the Food Safety Act 1990 on the administration of medicines to food animals. These restrictions are comprehensive and take into account a number of options which apply to the type of products that can be used for treatment of a particular condition in the animal species under consideration. Legal options are available for (1) the use of a product that is fully licensed for the animal species and condition under treatment; (2) the use of a product licensed for another condition in the same species; (3) a product licensed for use in another animal species; (4) a product licensed for human administration; (5) the use of special products made up under a veterinary prescription by a person authorised for this purpose. The first option is the one of choice with the others only being used under exceptional circumstances. With options 3, 4 and 5 only a small number of animals on a specified farm may be treated in this manner. With options 2,3,4 and 5 the product administered may contain only substances to be found in a product licensed for food animal species; administration must be by a veterinary surgeon to whose care the animals are committed or be in accordance with his directions but his presence is not mandatory; adequate records must be kept by the veterinary surgeon as follows; (i) date of examination of the animals; (ii) details of their owner; (iii) diagnosis; (iv) products and dosages administered; (v) number of animals treated; (vi) duration of treatment and withdrawal period recommended. These records must be kept and be available for inspection for a period of not less than three years.

### Injectables

The injection of an antibiotic into an animal ensures that the required amount of the medication is with certainty given to that animal. To ensure complete biological availability of the antibiotic it must be injected intravenously, intramuscularly, intramammary or subcutaneously. Other routes such as intraperitoneally are less frequently used.   If the antibiotic is injected into fatty deposits then most of the antibiotic will be unavailable to the animal because of the poor blood supply  and binding of the antibiotic to the fatty deposits. The injection of a single farm animal does not usually present any difficulty other than the requirement for a degree of restraint prior to handling. The same however, cannot be said for large numbers of animals especially flocks of poultry which may require penning or housing, both of which could result in distress and welfare problems for the animals concerned. The use of hypodermic needles on farms is not always a safe procedure and there is a serious risk of causing an abscess at the site of injection especially if the needle becomes contaminated during use.

Many comments have been made that antibiotics should not be used at all in animal production. Such statements show a complete lack of understanding of animal disease and of how disease-causing-agents are maintained and transmitted in nature to animals and animal populations. These statements also show an inhumane disregard for animal welfare because, as mentioned above, many diseases cause the affected animals severe pain or serious discomfort and if these animals were to be left untreated, then they would remain in pain until they either died or recovered. Death may also be a consequence of many infections and this, together with the extreme stresses of infectious disease, can be prevented by judicious antibiotic use. No livestock farmer, no animal owner and no veterinary surgeon would accept a situation which prevented sick animals from being treated with medication to prevent pain and to save lives, apart from reducing the dramatic economic losses associated with animal disease.

## (a) Acute clinical disease

The most common way of introducing a new disease or a new variant of an 'old' disease onto a livestock unit is by introduction of live animals. Acute bacterial disease on a unit is frequently associated with either management failure, a change in the climatic environment or an intake of infected animals. In all of these cases, antibiotics may be required to treat the infected animals and prevent the disease from spreading. A veterinary surgeon will normally be involved in the diagnosis of the disease outbreak and also in prescribing the antibiotic for the sick animals under his care

Farm animal and fish veterinarians are aware of the need to protect the health of the human population and will make every effort to use antibiotics only when necessary and for the least possible time.  There are many statutory restrictions on the use of antibiotics in clinically sick food producing animals and violations attract serious penalties both for the farmer and his veterinarian. The antibiotic has to be used according to the data sheet regulations, the treated animal or group of animals must be clearly identified and the correct withdrawal period or interval between use of antibiotics and slaughter must be observed. Records have to be kept in respect of all animals that are being treated; these animals must be clearly recognised as receiving treatment and the completed records have to be available for inspection by officials. Also, as mentioned above, surveillance does not end at this point because EC regulations demand that steps are taken to monitor meat at the slaughter house for residues of medicinal products that might have been used in meat production. If carcases are detected with tissue residues of antimicrobials greater than the maximum residue level (MRL) permitted by the regulatory authorities, they will be condemned as unfit for human consumption and the farm of origin will be visited by officials in an attempt to determine precisely how the residue violation occurred. There are also provisons for financial penalties if violations are detected. Further discussion on MRL's will be covered in the section on consumer protection. It is clear therefore that the veterinary surgeon has to take into account a whole variety of factors relating to legislation and human safety before he decides to medicate a clinically ill food producing animal.

### (b) Low grade chronic disease

Several types of pathogenic bacteria can infect poultry, pigs and cattle which may then only show a slight drop in production or growth rate rather than any specific clinical sign of disease. The recognition of the presence of these almost inapparent infections is usually based on reduced performance because clinical signs are not seen and death loss is often not recognised. In these circumstances, post mortem examinations are generally not carried out in sufficient numbers to confirm the presence of a specific disease-causing-agent. Under these conditions and with a tentative diagnosis, the use of an appropriate therapeutic antibiotic has often been shown to return the herd or flock to previous levels of economic performance. It is not always possible to know the identity of a bacterial agent causing disease in a sick patient especially if samples cannot be taken for microbiological examination. The use of antimicrobials in this way is entirely cost effective and, with the application of relevant safety measures, this presents no risk to the consumer. Any sick or treated animal will not be sent for slaughter until it has fully recovered from its illness and a withdrawal period appropriate to the antimicrobial has been applied. It must be made absolutely clear at this stage that diseased animals are never accepted into slaughterhouses because they will be recognised as such during mandatory ante mortem inspections.

*Large farm units*

The livestock farming industry has been encouraged by successive Governments, sometimes with the aid of subsidies, both hidden and transparent, to develop intensive methods of farming as a means of producing an ever increasing quantity of food at a realistic price for an expanding population. Livestock agriculture achieved this change with a high degree of efficiency with the cooperation of geneticists, who selected animals for prolificacy and productivity, and also with nutritionists, who formulated diets specifically for these high yielding hybrid animals. Initially, these improvements were made with poultry, but they soon extended to the pig and latterly they have encompassed both freshwater and saltwater fish and also crustaceans. Completely new methods of husbandry were required to rear and grow very large numbers of intensively farmed food-producing animals. New techniques had to be developed in order to safeguard animals housed indoors and also to manage them in the most cost effective manner. Cages, batteries, stalls and tethers were introduced to enable large numbers of animals to be managed as efficiently as possible and also to prevent the bullying, and fighting that always occurs when certain types of animals are kept in direct contact with each other. These intensive housing systems are currently under scrutiny because some of them are considered to provide low standards of welfare. The intensive housing systems have gradually produced a very docile animal which in itself has resulted in better animal performance because of a lack of stress to the animal from the immediate surroundings. Highly digestible food is provided often several times a day, the climatic temperature is kept constant, there are no predators and management provides regular working routines. Currently there are no widespread programmes to select animals that are genetically resistant to the common diseases. This is due in part to the poor heritability of these characteristics but also because breeding programmes are being devoted to productivity traits which are perceived as being more economically viable than resistance to disease. Contrary to commonly held opinion, intensively reared animals are no more susceptible to infectious disease than any other animal kept indoors or outdoors. In fact, many of the pigs and poultry that are kept in controlled environment units suffer from fewer disease outbreaks than conventionally reared animals. This is because the indoor animals are out of contact with the soil and many kinds of disease-carrying wild rodents and birds and also have no contact with other infected pigs or poultry due to the operation of security barriers and movement restrictions on new animals entering the unit. Similar restrictions apply to the movement of humans entering these units who must be free of pig contact for at least 48 hours and must put on a complete change of clothing after taking a shower and before entering the pig unit.

As mentioned above, indoor housed pigs and poultry can still be infected by a variety of disease-causing agents. It is the task of management to recognise these infections as early as possible and prevent them spreading around the herd or flock by administering appropriate medication. As discussed in the section on legislation, all owners of food producing animals have a legal obligation to ensure that when their animals are sent for slaughter they do not contain in their tissues any residues of medicinal substances at levels above the Maximum Residue Limit. Also regular monitoring for residues will be carried out by the Veterinary Medicines Directorate and all violations will be investigated. When an animal is treated for a bacterial disease then, by law, a record must be kept about the type and amount of antibiotic that is used, the withdrawal period that must be applied and the identification of the animal concerned. Veterinary surgeons when visiting the unit will expect to view these records which will also be checked by visiting MAFF and RPSGB officials. So, a record of antibiotic use on farms will be available when investigations are being carried out concerning the presence of residues of medicines in meats and meat products.

It is a widely held view that antibiotics are overused on food producing livestock units. In practical terms this is not the case for several reasons, the most important being that antibiotics are relatively expensive substances and farmers will not use them unless they are necessary. Secondly, farmers are generally unwilling to use antibiotics constantly on a widespread basis because of the risk of residue violation. Veterinary surgeons also are not willing to prescribe antibiotics unless they are absolutely necessary; however, they are also conscious of the welfare and economic benefits of both strategic and herd medication within intensive systems.

The situation with respect to cows' milk and hens' eggs is very similar to that for pig and poultry meat. Antibiotics that are used for the treatment of dairy cows are also subject to very stringent legal restrictions backed up by regular monitoring of milk and financial penalties for violations. Similar restrictions also apply to egg and broiler production with specific or standard withdrawal periods being provided for products used to treat bacterial disease in laying hens or broiler chickens. The situation is similar for farmed fish and crustaceans and specific and standard withdrawal periods are provided for the therapeutic antibiotics used to treat bacterial disease.

As has been clearly proven with cow's milk, the best method of reducing residue violations in meat, eggs and fish is to introduce regular monitoring and levy strict financial penalties on those producers who persistently supply foodstuffs for processing that contain illegal levels of residues of medicinal substances.

### (c)   *Management considerations*

Continued intensification of farm animal production has indicated a specific need for changes in attitudes towards established agricultural practices. These changes have been driven by a need to increase production efficiency and to reduce costs by using healthier livestock so that fewer diseases would be experienced and thus there would be less need for antibiotic medication. Unfortunately, this has meant undue attention being paid to the cost/efficiency of production instead of the animal or product that is being produced. The overall health of pigs and chickens has increased markedly over the past ten years, mainly because there has been greater use of minimal disease breeding stock. This has been brought about by the larger breeding companies buying up the smaller units and buying and selling breeding replacements that have been produced and reared free from specific diseases. Such improvements have enabled many pig and poultry rearing units to purchase and use minimal disease animals and one advantage of this has been to reduce the spread of infected animals around the country. Obviously, the maintenance of a minimal disease status can not be achieved without improvements in other areas such as management skills and environmental control. Most of the larger animal breeding and rearing organisations provide training for their animal staff on a wide variety of subjects including health control, early disease recognition and the proper use of medicines to avoid residues in carcases at slaughter. A further development in animal health, which has reduced the need for antibiotic medication, is the production of new and improved vaccines. Many efficient vaccines now control diseases which in the past have tended to cause serious health problems. This has provided efficient immunological methods to control many disease conditions which hitherto would have necessitated prolonged antibiotic medication. Animals that have been produced and reared free of specific diseases are still susceptible to a wide range of ordinary bacterial infections for which there is no vaccine available. Often these infections affect only one or two animals and appropriate injectable medication will very often effect a total cure. There are, on the other hand, several bacterial diseases which can rapidly spread through a herd or flock and, unless all animals at risk including the diseased ones are medicated with an appropriate antibiotic at one and the same time, the disease will continue to spread and many animals will die. It is essential to medicate in this way because the animals are usually living in the same building, drinking the same water and breathing a common air supply. It is highly likely, therefore, that many of the animals will have been infected at the same time and it is only prudent to treat them at the same time to curtail the disease and stop pain and suffering. A similar requirement for antibiotic medication follows viral infections when, due to damaged tissues or lowered immunity levels, secondary bacterial infection occurs. Supportive antibiotic therapy ensures that death losses are kept to a minimum and secondary bacterial infections do not occur. The use of antibiotics in this manner is an essential method of disease control in large herds and, if used speedily and with due concern for duration and level of dosage, the majority of epidemic bacterial disease will be halted with minimal death loss.  Such methods of medication are well known to military medicine and disease control programmes in undeveloped countries in the face of epidemics or when natural disasters have occurred and serious outbreaks of human disease are threatened.


PERFORMANCE/EFFICIENCY ENHANCERS ( GROWTH PROMOTERS)

Since antibiotics were first shown to bring about an improvement in animal performance in the late 1940's, much work has been carried out in an attempt to understand the mechanisms of action of these substances on the intestinal bacterial flora. This work was interrupted following publication of the Swann report (1968) because it prohibited the use of therapeutic antibiotics such as the tetracyclines, penicillin, streptomycin and the sulphonamides as growth promoting agents in food producing animals. Other substances such as Zinc Bacitracin were, however, soon recognised as being capable of bringing about the same degree of performance improvement as the therapeutic antibiotics and were accorded the classification of Feed Antibiotics by the Swann Committee. Research was then continued on the modes of action of these new substances and this introduced a new dimension into the investigations because unlike the therapeutics, the new feed antibiotics were not absorbed through the intestinal tract into the body tissues. This meant that residues would not occur and that feed antibiotics could be fed right up to the point of slaughter without presenting any problem for human health.  The current theory  about the modes of action of these feed antibiotics is that they prevent wide fluctuations in bacterial population numbers  within the intestines and they also inhibit the formation of toxic substances that are produced as part of normal bacterial metabolism. Other activities also occur in that various intestinal bacteria, including *E.coli* and Salmonellae, become more sensitive to the action of therapeutic antibiotics after having been in contact with these performance enhancers. This may result from a two fold action. On one hand, there will be a lack of challenge by a therapeutic antibiotic causing a gradual return to full susceptibilty of the bacteria.  There will also be a direct effect of the performance enhancing substance on the gut bacteria themselves causing resistance genes to be lost (curing) with a consequent return to a non-resistant state. It is also clear that performance enhancers do not cause bacteria to become resistant to them because after almost forty years of agricultural use they are still highly effective and continue to produce the amount of performance improvement as when they were first introduced as a management tool. There is some experimental information that continuous administration of these performance improvers at a very low level may help to prevent the development of some  diseases, especially in pigs.

The management methods for very young livestock vary greatly between pig and poultry units. This is related specifically to the ability of newly hatched chickens to be self sufficient from the first day of life. Piglets on the other hand, need to obtain milk from their mother in order to acquire protective antibodies. Also the piglets' intestinal tract at birth is only able to digest milk containing a simple sugar eg. lactose and other very readily digestible nutrients. Newly hatched chickens can therefore be transported over long distances to a rearing and growing unit because their body contains a good supply of nutrients in the form of egg yolk which will last them for up to 72 hours. Piglets on the other hand, must remain with their mother for many days before weaning to get the benefit of a good milk supply, to develop a mature digestive system and to acquire immunity against a variety of infections. The skill of the manager aims to prevent both the young chicken and piglet from picking up a variety of diseases during this period and he is aided in this by scientific use of a variety of management aids including disinfectants and antimicrobials.

Many supermarket chains are now complicating management systems by laying down requirements, sometimes unreasonable, for the husbandry and management of the food animals that they intend to purchase and sell through their retail outlets. These requirements concentrate especially on welfare and on trying to reduce the use of in-feed antimicrobial substances including performance enhancers. The supermarkets are trying to ensure that any meat, eggs or milk that they sell will be produced in a way acceptable to the consumer. Also, they consider that because of the controlled methods of production, these foods will not contain any antimicrobial substances which would place the consumer at risk. Obviously by doing this the supermarkets also hope to improve their image and benefit their sales. However, it must be evident that most systems of animal production that incorporate judicial use of antimicrobials can maintain output at a reasonable cost which can be passed on to the supermarket and thence to the consumer.

## Introduction of new methodology

Two new technological advances have been introduced into animal agriculture over the past decade and these, in the long term, may have some effect on the overall level of antibiotic being used, especially in pigs and poultry.

1) Probiotics ('micro-organisms'). These are preparations consisting of live or dead bacteria or bacterial spores and, in most cases, are given to animals as an alternative to antimicrobials for the purpose of enhancing performance. Probiotics were introduced because it was considered that their use would not result in tissue residues and that there would be no problems similar to

antibiotic resistance and cross resistance. Initially, it was suggested that probiotics could totally replace antimicrobials in every respect. It soon became very clear, however, that this was not the case and that, whilst the end result of their administration to animals may in some respects be similar, the means of achieving this result was very different for the two types of product. Some of the probiotics will not give rise to residues but others will, during multiplication in the intestines, elaborate antimicrobial substances that will be absorbed from the gastro-intestinal tract and may lead to the production of residues in edible tissues. These residues would be detected at the slaughterhouse and may lead to condemnation of the carcass. The mode of action of the probiotics, presently a field of intensive research, is not very clear but it has been suggested that some of them may work by generating an acidic environment that will not permit certain bacteria to grow. Alternatively, it is thought that some probiotics colonise large areas of the intestinal mucosa and simply by competitive exclusion prevent the growth of unwanted bacteria. Probiotics may also compete for limited nutrients in the intestinal tract and so help to prevent excessive multiplication by potential toxin producing organisms. Many claims have been made in respect of probiotic activity and product composition. Statutory requirements are currently being prepared by EC and these might generate an insight into the modes of action of the probiotics.

2)    Enzymes. These are being given in the feed to food producing animals in order to influence productivity. Lipases, phytases and cellulases in particular appear to be of value by releasing nutrients from low grade poor quality foods that would be otherwise unavailable to the host animal. The provision of extraneous enzymes will therefore help maintain growth and productivity. As enzymes are protein in nature, their degradation in the intestinal tract will lead to the production of amino acids, some of which will be utilised by the animal during digestion.

Problems are presented with probiotics and enzymes during processing of feed, especially for pellet production. At present, there are no statutory requirements in respect of providing data on quality, safety and efficacy for probiotics but a directive recently introduced by the EC will provide for the review of data especially on safety but it may also yield an insight into their modes of action. Products containing bacteria have been administered to day-old chickens in an attempt to prevent the establishment of salmonellas in the intestines which at hatching are devoid of a bacterial flora and this technique has been reasonably successful. This procedure was initiated in an attempt to control or reduce the prevalence of salmonellosis in human populations. However, in those countries that have introduced the technique, there does not seem to be any overall diminution of salmonellosis in the processing of feed, especially for pellet production. This type of processing uses high temperature and pressures, both of which are detrimental to the viability and stability of probiotics and enzymes respectively. The increasing use of extruders, which heat the meal to very high temperatures (125°C) for a very short time (1-5 seconds) is likely to be a problem when including probiotics and enzymes.

## Consumer protection

Once a new antimicrobial substance is considered for use in animals, the manufacturer has to test it according to EC Directives, Regulations and Guidelines. According to the Medicines Act of 1968 all medicinal products have to undergo a series of tests in order to prove that they are not going to cause any hazard for human consumers or for the environment. A series of criteria relating to quality, safety and efficacy must be satisfied and a Maximum Residue Limit (MRL) must be established before a licence for use is issued. Once a medicinal product is being used in the field a further safety measure is brought into play. A veterinary surgeon suspecting that an adverse reaction has occurred as a result of administering the product is obliged to submit a full report of the suspect adverse reaction to the VETERINARY MEDICINES DIRECTORATE (VMD). This report will be considered together with any others for the same product and the VMD may require further testing of the product; it may introduce new guidelines for use or, exceptionally, the licence for the product may be withdrawn on safety grounds. All the above testing procedures together with the residue monitoring being carried out ensures that medicinal products being used in animals are as safe as they possibly can be when used according to the data sheet instructions.

Probably the most effective consumer protection measures relate to quality assurance schemes that have been introduced both by the Ministry of Agriculture and the integrated supermarkets. The Ministry of Agriculture has introduced such a scheme aimed at the pigmeat producer termed the Quality Assurance Scheme. Pig farmers joining the scheme receive a minimum of four veterinary visits each year, three by the local veterinarian and one by a Ministry vet. During these visits welfare considerations will be a high priority, documentation related to overall antibiotic usage will be scrutinised and samples will be taken to check on the possibility of salmonellas being present in the faeces. Many supermarkets set out a series of guidelines stating the conditions under which livestock should be reared, when medicines are permitted and also carry out their own routine inspections of producers premises.

PUBLIC HEALTH: STATUTORY CONTROLS

### The producer of medicated feeds

The statutory basis for regulating the manufacture, sale and supply of all veterinary medicinal products is provided for under the The Medicines Act 1968. The Medicines (Medicated Animal Feedingstuffs) Regulations 1992 require that the incorporation of medicinal products in feeds should only be undertaken by registered manufacturers. The regulations also prohibit the sale, supply or import of animal feedingstuffs containing a prescription only medicine unless it has been incorporated in accordance with a veterinary written direction, this being an instruction by a veterinarian to a registered feed manufacturer to incorporate a medicinal product into an animal feedingstuff.

Stringent legislation therefore is being applied to ensure that there are no risks to man from the use of antimicrobials in animal feeds. Animal feedingstuffs containing antimicrobial substances can only be prepared, sold, imported or supplied in accordance with specific legislation which includes a product licence, animal test certificate (ATC) or veterinary written direction (VWD). Manufacturers of medicated feeding stuffs must register with the Royal Pharmaceutical Society of Great Britain (RPSGB), or the Department of Agriculture for Northern Ireland. This means that both the large commercial feedingstuffs manufacturer (Category A mixer) and the farmer who mixes his own pig or poultry feed (Category B mixer), must, as a legal requirement, register with the RPSGB. Furthermore, any medicines, such as antimicrobials, that are mixed into feedingstuffs must be licensed feed additives and incorporated in accordance with the data sheet recommendations. These stringent controls mean that the livestock farmer can use medicated feeds with great confidence in the knowledge that they have been prepared to the highest specification in respect of quality of the antimicrobial being used, and that all safety measures have been applied to ensure that no adverse reaction will occur if the medicated feed is used in accordance with the manufacturers instructions and the relevant data sheet recommendations.

### The user of medicated feeds

The general public is rightly concerned that the most rigid requirements are met in relation to the safety of medicines, especially antimicrobials, that are used in the production of meat and other animal products for human consumption. Also, the pharmaceutical manufacturers and veterinary surgeons, who are equally concerned about safety, are anxious that sufficient efficacious medicinal products are available to treat the range of infections and diseases that are associated with food animal production. The UK Veterinary Medicines Directorate (VMD), established as a result of the Food Safety Act (1990), is responsible for recommending acceptance or refusal of licence applications and ensuring that EC regulations are met in respect of all antimicrobial substances that are used or being proposed for use in food animal production. This also applies to medicinal products that are imported into UK from other countries. The VMD also administers the Veterinary Medicines Residue Testing Scheme and the Suspected Adverse Reaction Surveillance Scheme. It is quite clear, therefore, that a comprehensive system is in operation to ensure that the consumer is well protected. The introduction of the Medicines Act (1968) and the succeeding array of legislation (amounting to approximately 200 documents), together with the results obtained through their application in the field, should convince the general public that everything that can be done is being done to ensure the safety of food produced from animals. There is concern, however, that a degree of 'overkill' is being applied by various authorities in respect of medicines legislation. Many medicinal products are being forced off the market for technical reasons associated with the interpretation and application of legislation, sometimes on so-called spurious safety grounds, even though similar products continue to be used in human medicine. It is essential that the needs of the sick animal are given priority over the niceties of legal interpretation and application. A realism must be introduced into product licensing, especially over combination products and those substances in the British Pharmacopoea which have a proven record of success in the field of disease control or that are being used at therapeutic dose levels in human medicine. The authorities also must be forced to do more than apply a piece of legislation simply because it exists. As mentioned above, the treatment and welfare of the sick animal demands paramount consideration and these two must take precedence over legalistic niceties.

The user of a medicated feed has a major role to play in the whole area of safe food production. One must ensure that animals receiving this feed are clearly identified and that the details of medication are recorded and the appropriate withdrawal period is stated in the medicines administration book. There are powers to detain animals on the farm to ensure that residues of medicinal substances fall to acceptable levels prior to slaughter. The recording of this information is required by law and the records must be available for inspection if required (The Animals and Fresh Meat 'Examination for Residues' Regulations 1988). In particular, the regulations require a record to be made within 72 hours - either of the administration of a veterinary medicinal product to an animal or of the slaughter of an animal. The record book must be completed for any treatment supplied or administered by a vet as well as those applied by the farmer. Once the disease problem has been resolved, the owner must ensure that the correct withdrawal period has been applied before the animals are sent for slaughter. It is also an offence to administer an unlicensed medicine to a food-producing animal (unless authorised by a veterinary surgeon). Any unauthorised use will result in such animals being condemned as unfit for human consumption. Failure to observe withdrawal periods and the sale for slaughter of animals containing residues in excess of the MRL are both offences under the above regulations.

### At the slaughter house

Food animals are not permitted entry to the slaughter house if the attending veterinary inspector during an ante-mortem inspection considers them unfit for human consumption. Once inside the slaughterhouse, the carcass may be sampled under EC Directive 86/469 for the presence of antimicrobial residues which, if found to be in excess of the MRL, will contravene the Food Safety Act (HMSO 1990). This Act applies not only at the slaughterhouse but also throughout the whole of the food preparation chain up to the point of sale to the general public. Following the identification of any medicinal residue above the MRL in fresh meats, penalties as described elsewhere may be levied on the producer of the meat who would be obliged to disclose information on the administration of antimicrobials and any other information on withdrawal periods.

### The food animal veterinarian

Farm animal and fish veterinarians are very aware of the need to protect the health of the human population and make every effort to use antibiotics only when necessary and for the shortest possible time. As mentioned previously, there are many legal restrictions on the use of antibiotics in food producing animals and violations will attract serious penalties both for the farmer and the veterinarian. The antibiotics have to be used according to the data sheet regulations, the treated animals must be identified and the correct withdrawal period observed. Completed records have to be available for inspection by officials. Also, as mentioned above, surveillance does not end at this point because EC regulations demand that steps are taken to test meats at the slaughter house for residues of medicinal products that might have been used in the production of these food animals. If carcases are detected with tissue residues of antimicrobials greater than the MRL they will be condemned as unfit for human consumption and the farm of origin will be visited by officials in an attempt to determine precisely how the residue violation occurred. Hence the reason why medication records have to be kept and treated animals identified.

It is clear from the preceeding discussion that serious efforts are made to ensure that edible tissues from farm animals are produced free from potentially hazardous residues. Consumers can be assured that the systems of monitoring the whole process of antimicrobial use in food producing animals are as comprehensive as possible in order to ensure the wholesomeness of the food they are eating.


## TISSUE RESIDUES AND THE SIGNIFICANCE FOR HUMAN HEALTH

The Maximum Residue Limit (MRL) is defined as being 'that amount of antibiotic that must not be exceeded in food destined for human consumption'. A 'no effect' level is determined from established criteria based upon a knowledge of the safety, toxicology and microbiological effect of the product. After the no effect level has been established, a safety factor (usually 100 or 1000 times) is applied to provide the value for the MRL.

Below this amount, it is suggested that human intestinal bacteria are not influenced in any way, but in excess of it the intestinal bacteria may be effected. However, this is not to imply that all such effects are detrimental to the host. Under the auspices of the WHO and FAO, an international committee has been established known as the Codex Alimentarius. This committee is charged with the responsibility for establishing MRL's which are accepted at the International level. In this way it should be possible to achieve a level of international harmonisation in the global trade of animal products. Following cessation of administration of an antibiotic to an animal, a specific period of time must elapse to enable the antibiotic to be cleared from the animals' tissues before it can be sent off for slaughter. This period is defined as the Withdrawal Period. When a medicinal product is granted a product licence then a defined withdrawal period will be assigned. This is a major issue in that every medicine for food animal use has a stated specific withdrawal period. Screening tests are available for use in slaughterhouses to detect antimicrobials that may be present as a residue in an animal carcass. All medicinal products are being reviewed and reassessed by the EC Committee for Veterinary Medicinal Products in order to establish a legally binding MRL. Such reviews are extremely costly and are being required even though a product is being used in human medicine in much greater quantities than would ever be present in an animal carcass. Such a draconian interpretation of the legislation is resulting in the withdrawal of many medicinal products from the veterinary formulary. It is suggested that certain medicines legislation is currently being applied and interpreted in such a way that the application and implementation of the legislation is becoming more important than the provision of a wide range of safe and efficacious medicines for the treatment of disease in sick animals. The successful use of legislation requires a high level of skill based upon realism and a pragmatism that is not always obvious in relation to antimicrobial substances.


ANTIBIOTIC RESISTANCE

One problem with using antimicrobial agents to treat bacterial infections is that prolonged use of these agents may select some bacteria that are resistant to them. This is an understandable phenomenon because in any large population of bacteria there will be some that, because of a random mutation, will resist the action of the antimicrobial. This type of bacterial resistance, sometimes called mutational resistance, is seen with streptomycin and with the older quinolones like nalidixic acid. The level of resistance selected is usually of a low order and can often be overcome by applying a higher concentration of antibiotic that selected the original resistant strains. A more common type of resistance in bacteria, especially those that inhabit the intestinal tract, is due to the presence of resistance plasmids that are readily transferable to other similar bacteria and confer immediate resistance to one or more antimicrobial agents. There is no real evidence to suggest that the use of antibiotics actually causes resistance to occur in bacteria, but these agents will select out from a bacterial population those bacteria that are already resistant and which in the presence of the antibiotic may develop into a dominant antibiotic resistant population. Once the antibiotic is withdrawn, the resistant population will regress until once again the dominant population will be antibiotic susceptible.

The use of antibiotics in farm animals was initially viewed with concern because of the potential to select resistant bacteria which could then transfer to humans and there interfere with antibiotic medication. Of greater concern was the possible presence of antibiotic residues in meat, milk and eggs which, once consumed by humans, would select bacteria resistant to the specific antibiotics in the intestinal tract. Over forty years of antibiotic use in humans and farm livestock has indicated quite clearly that neither of these two scenarios present a realistic hazard for man, a point that needs to be heeded by alarmist naive scientists and journalists alike. Most animal bacteria with few exceptions are not able to colonise sites within the human body, a procedure that must occur before resistance plasmids can transfer to other indigenous bacteria.   '


HAZARD/RISK RELATIONSHIP OF ANTIBIOTIC RESIDUES FOR MAN

There have been many reports suggesting that some human disorders may have been caused by antibiotic residues in the diet. The concentration of any antibiotic residue, even when accumulated across all foodstuffs, would be far too low for any direct toxic reaction to occur to the consumer. If any effect were to be produced by antibiotic residues, it would essentially be directed against the intestinal bacteria and for any such effect to occur there would have to be a threshold level of residue present which would inhibit bacterial growth, possibly alter its morphology and in very exceptional circumstances, actually destroy specific groups of bacteria. In order for such extra-ordinary levels of residues to be present, the animals would have had to have been treated with quantities of antibiotics that would be exceptional, illegal and wholly unrealistic in pharmacological or even in financial terms. It is often suggested that immunological processes are involved in some human disorders, but often because other mechanisms have been ruled out and very rarely because the nature of the clinical disorders or laboratory tests give a positive indication of such reactions. Another hypothesis often propounded is the possibility of idiosyncratic responses in individuals to minute amounts of antibacterials in the diet. This is useful to some activists because of its vagueness and unprovability but which in reality does not lend itself to scientific assessment and is impossible to reproduce under controlled conditions. Hence the claims of "antibiotic induced food intolerance" in ordinary members of the general public are almost always vague and uncertain, they lack adequate supporting data from laboratory tests or controlled challenges and they do not fit known patterns of disorders and mechanisms. For this reason, antibiotic contamination of food as a notional hazard has not been shown to present a practical risk.

A further postulated hazard relates to the possibility of antimicrobial residues being present in meat, milk, eggs or fish which are consumed by humans as part of their diet. There is a special contractual agreement with milk producers in England and Wales for all milk to be discarded for the first 4 days ie 96 hours after the cow has calved, irrespective of any product withdrawal periods which in fact may be less than this. In addition, in Scotland it is a requirement that milk is witheld for a minimum of 48 hours after the administration of any medicine, again regardless of any lower recommended withdrawal period. There are also special withdrawal conditions for dry-cow therapy and also for use in the event of premature calvings. Milk has a very special place in the human diet particularly because it is widely used for children and is considered to be the premier natural health food. The regulations quoted above are designed therefore to protect this valuable source of nutrients and especially to safeguard the manufacture of cheese and yoghurt. Many varied and complex interactions on the intestinal bacterial flora have been suggested. In many cases, the precise concentration of the residue has not been taken into account, neither has the cooking process nor the diluting effect of the large volume of the intestinal contents nor its ability to destroy antimicrobial substances been considered. With reference to the selection of antibiotic resistant bacteria, there is a critical concentration of antimicrobial that must be present in order to select resistant strains of bacteria and this is referred to as the Minimum Inhibitory Concentration (MIC). This is also defined as the lowest concentration of antibiotic that will inhibit the growth of bacteria. If bacteria survive and also multiply in this concentration, then they are generally considered to be resistant. The majority of MIC determinations have been carried out in-vitro under specific experimental conditions. The values that have been obtained may have little relevance for the in-vivo situation in which antibiotics have to show activity in the presence of tissue fluids, blood constituents and in some cases other micro-organisms. MIC values are arbitrary values only and, whilst they can be compared with obtained antibiotic blood concentrations, they must be recognised for what they are and that is laboratory derived data only. It is of major concern that some regulators are considering the use of MIC values obtained in highly artificial experimental animal model systems, which are wholly incapable of realistic scientific assessment, as a means of generating acceptable MRL's. This is an example of extrapolation at its most ludicrous. Many surveys have shown that antibiotic residues are relatively rare in human food and that the concentrations detected are very low, often less than one part per million. This minute amount of antimicrobial may be affected by the cooking process, and if the substance is heat labile it may be totally destroyed. Even if the substance is heat stable, it may also be destroyed by the action of the different food ingredients with which it is combined. The experimental data on this aspect of antimicrobial destruction and/or survivability is very sparse indeed. Once a complete meal is produced and ingested there will be a significant dilution, especially by the volume of the intestinal contents, such that any antibiotic that had been present in the dietary ingredients would probably be totally destroyed. On the other hand any remaining antibiotic would be so small in amount that it would be unable to achieve an MIC and thus would be impossible to select any resistant strains from the intestinal bacterial contents. Up to the present time, no data are available which indicate that antibiotic residues in food select resistance in human intestinal microflora. Reports from the National Residue Surveillance scheme indicate that during the period January to December 1992 over 31,000 samples were tested from cattle, sheep and pigs and less than half of one per cent contained any detectable antimicrobial residues (National Surveillance Scheme for Residues in Meat. Results reported in MAVIS : Medicines Act Veterinary Information Service, issued by the Veterinary Medicines Directorate, Woodham Lane, New Haw, Addlestone, KT15 3NB. UK). In January 1992, much stricter controls were introduced in UK to prevent unacceptable residues of veterinary medicines occurring in meat. This control is regulated by the 'Animals, Meat and Meat Products (Examination for Residues and Maximum Residue Limits) Regulations 1991, Statutory Instrument No. 2843[1]. Although Britain has a very comprehensive system of residue monitoring with sampling at all stages from the farm to the shop, the new controls were considered necessary to finally reassure consumers that harmful residues of all types of animal medicines, including antibiotics, are not being transmitted through the food chain. The new regulations make it an offence to use unauthorised drugs in food producing animals; they enable animal carcases found to contain residues of illegal drugs, or residues of authorised animal medicines in excess of certain prescribed maximum residue limits (MRL) to be condemned for human consumption. Also the regulations make it an offence to sell for slaughter animals containing unauthorised drugs or medicinal residues in excess of the prescribed MRL's and also for shops to sell meat containing such residues. The regulations require farmers using authorised medicines to observe specified withdrawal periods prior to slaughter and they give powers to officials to detain and check animals or meat if misuse of animal medicines is suspected.

Of special concern are the figures for sulphadimidine residues. Basically, this antibacterial substance has a very simple chemical structure but, because of the ionic charges on the surface of the molecule, it has a particular affinity for adherence to the surface of most materials. It can, for instance, remain in sufficient concentration on the surface of a washed concrete floor to cause the next batch of pigs admitted to that pen to take in by mouth enough of the residual sulphonamide to exceed the MRL in the tissues. This electrostatic effect is of great concern to the farming industry to such an extent that an increasing number of veterinarians and farmers are refusing to use this antimicrobial for animal treatment. It has to be stressed the substance still has wide application in the control of animals health and this can continue with correct usage of the product, avoidance of carry over in faeces and using suitable forms of product such as granules rather than the powder form.

The finding of antimicrobial residues in animal tissues and the subsequent official investigation does not always lead to a clear-cut explanation as to their presence. Sometimes there has been confusion during mixing of the food, sometimes the wrong food has been delivered but most disturbing of all is when after several detected MRL violations, no reason has been found to explain the presence of the antimicrobial tissue residue. Such findings are inevitable with extremely sensitive laboratory analytical procedures and with antibiotics that are naturally produced in the animals' environment. It will only be by careful sampling of tissues, in depth epidemiological study, by cautious standardisation and application of realistic laboratory methodology that such anomalies will be unravelled.

## Conclusions

Experience obtained over forty years of antimicrobial use in animal production has clearly indicated a valuable role both for the treatment of disease and the improvement of animal performance. One unexpected finding has been the continued efficacy of some of the earliest antimicrobial agents such as penicillin and the tetracyclines after almost fifty years of continuous use. This is in part due to the rapid turnover of the different animal populations but it can also be explained by the fact that resistance to antibiotics in bacteria is not an absolute phenomenon. It is now also clear that tens of millions of food producing animals never receive any therapeutic antibiotic at all during their short lifespan.

The regulations that are in operation to safeguard human health are somewhat of a paradox because little to no real evidence exists that the use of antibiotics in food animals is detrimental to human health. It is highly likely that the so-called hazards of antibiotic use in animals have been confused and related by many people to the whole subject of bacterial contamination of human food and poor standards of food hygiene. Human food poisoning bears no relationship at all to the use of antibiotics in animals, despite many media statements to the contrary.

It is clearly accepted that some form of regulation is required for the use of antimicrobials in livestock production in order to protect both animal and human health. However, unless a more pragmatic approach is taken by legislators to the proven safety record and lack of hazard of agricultural antibiotics, then the whole animal industry runs the risk of wholly unwarranted, grossly expensive and mis-directed over-regulation. Inevitably this will lead to the withdrawal of numerous, therapeutically valuable products, the responsibility for which must rest entirely with the relevant licensing authority and its often overzealous interpretation of community legislation. These bodies must not lose sight of one of their main responsibilities in this area. These are to ensure the continued availability of safe, efficacious and high quality products needed to maintain the health and welfare of our domestic and food animal species.

## References

Food Safety Act (1990) H.M.S.O.

MAVIS (Medicines Act Veterinary Information Service) Veterinary Medicines Directorate, Woodham Lane, New Haw, Addlestone, KT15 3ND, UK

Medicines Act (1968) H.M.S.O.

Swann Report (1969) Report on the Joint Committee on the use of antibiotics in Animal Husbandry and Veterinary Medicine. H.M.S.O.

The Animals and Fresh Meat (Examination for Residues) Regulations (1988) H.M.S.O.

The Animals, Meat, Meat products (Examination for Residues and Maximum Residue Limits) Regulations 1991. H.M.S.O.

The Medicines (Medicated Animal Feedingstuffs) Regulations (1992) HMSO.

Walton, J. R. (1983). Antibiotics, Animals, Meat and Milk. *Zentrallblatt fur Veterinarmedisin.* A. **30**, 81–92.

# 3

# PUBLIC HEALTH PROBLEMS ASSOCIATED WITH THE USE OF ANTIBIOTICS

E. J. THRELFALL, J. A. FROST and B. ROWE
*Public Health Laboratory Service, Laboratory of Enteric Pathogens,
61 Colindale Avenue, London NW9 5HT, UK*

## Introduction

The question of whether antibiotic resistance in non-typhoidal salmonellas associated with food animals in the UK is a human public health problem has been a contentious issue since the late 1960s. For an understanding of the issues involved, it is important to be aware of the situation in the UK at the present time and of the factors, both past and present, which have contributed to the occurence and spread of antibiotic-resistant strains in food animals. It is also important to be aware of the measures which have been introduced to limit the appearance and spread of such strains in the UK.

## Occurence of resistance

*1965 - 74*

Following widespread concern about the increasing incidence of multi-resistant (= resistant to four or more antimicrobials) strains of *Salmonella*, particularly *Salmonella typhimurium* definitive phage type (DT) 29 (= DT 29) in humans and food animals, especially bovine animals (Anderson, 1968), the Joint Committee on the Use of Antibiotics in Animal Husbandry and Veterinary Medicine (the Swann Committee) recommended in 1969 that certain therapeutic antibiotics, at that time widely used in food animals without prescription, should be available only on prescription (Anonymous, 1969). A further recommendation of the Swann Committee was that certain antibiotics should be reserved specifically for prophylaxis and therapy, and should not be used for growth promotion. By 1970,

DT 29 had disappeared from bovine animals in Britain and for the next six years less than 8% of strains from cattle and 3% of strains from humans were multi-resistant (Rowe and Threlfall, 1984). Isolations of multi-resistant DT 29 were at a low level before the enactment of the Swann recommendations and factors other than restrictions on the use of antibiotics as growth promotors probably contributed to the disappearance of this multi-resistant strain.

*1975 - 1990*

From 1975 to the mid 1980s there was again a substantial upsurge in the incidence of multi-resistant *S. typhimurium* in humans and food animals, particularly calves. The phage types involved were different from those observed in the 1960s, with the related phage types DTs 204, 193 and 204c predominating (Threlfall *et al.*, 1978a, b; 1980; 1985). A feature of this outbreak was the sequential acquisition of plasmids and transposons coding for a wide range of antimicrobials - ampicillin (A), chloramphenicol (C), gentamicin (G), kanamycin (K), streptomycin (S), sulphonamides (Su), tetracyclines (T) and trimethoprim (Tm). The acquisition of resistance by strains of these phage types seemed to coincide with the introduction and use, in calf husbandry, of at least some of the antimicrobials involved, in attempts to combat infections with *S. typhimurium* resistant to an increasing range of antibiotics (for review, see Rowe and Threlfall, 1984). In particular, this epidemic provided the first conclusive evidence of the introduction and use in calf husbandry of a veterinary antibiotic (apramycin), giving rise to resistance to gentamicin, an antibiotic which is used for treating severe systemic infections in humans (Threlfall *et al.*, 1985, 1986). Following these observations, and taking into account the observation that both drug resistance and multiple resistance was continuing to rise in *S. typhimurium* from humans and bovine animals in England and Wales - in humans, from 5% in 1981 to 12 % in 1988 (Ward *et al.*, 1990; ) and in cattle, from 15% in 1981 to 66% in 1990 (Threlfall *et al.*, 1992, 1993), in 1990 the Expert Group on Animal Feedingstuffs (The Lamming Committee) recommended that not only should antibiotics giving cross-resistance to those used in human medicine not be used as growth promotors, but that their prophylactic use in animals be reconsidered (Anonymous, 1992). The Government response to this recommendation was that the advice of the Veterinary Products Committee (VPC) and the Committee on the Safety of Medicines be sought.

THE CURRENT SITUATION

Regrettably, since 1991 there has been a further substantial increase in the incidence of both resistance and multiple resistance in *S. typhimurium* isolated from humans in England and Wales and in 1994 78% of isolates from humans were resistant to at least one antimicrobial and 62% were multi-resistant (Frost *et al.*, 1995). The situation has not improved in 1995, and in the first 10 months of that year 83% of isolates were drug-resistant and 69% multi-resistant (Table 3.1). We are no longer able to provide information on the situation in bovine animals as since 1993, such strains have been both phage-typed and screened for drug resistance at the Central Veterinary Laboratory (CVL).

**Table 3.1** DRUG RESISTANCE IN *S. TYPHIMURIUM* FROM HUMANS AND BOVINE ANIMALS IN ENGLAND AND WALES, 1981 - 1995 (TO OCTOBER 30)

| | *Humans* | | | *Bovines** | | |
|---|---|---|---|---|---|---|
| | Total | % DR | % MR | Total | % DR | % MR |
| 1981 | 3992 | 36 | 6 | 1157 | 71 | 15 |
| 1990 | 5451 | 54 | 18 | 1178 | 79 | 66 |
| 1994 | 5603 | 78 | 62 | | | |
| 1995 (to Oct 30) | 5663 | 83 | 69 | | | |

Source: Laboratory of Enteric Pathogens data (figures for the first 10 months of 1995 are provisional)

DR, drug-resistant (to one or more antimicrobial);
MR, multi-resistant (to 4 or more antimicrobials)
* From 1993 bovine isolations have been screened for resistance by CVL (see text)

Of particular importance in this increase in the incidence of multi-resistance in *S. typhimurium* since 1991 has been an epidemic of multiresistant *S. typhimurium* DT 104 of R-type ACSSuT (resistance to ampicillin, chloramphenicol, streptomycin, sulphonamides and tetracyclines) in bovines and humans in England and Wales. This phage type/R-type combination was first identified in humans in 1984. Before 1988 less than 50 isolates per annum of DT 104 R-type ACSSuT were identified by the Laboratory of Enteric Pathogens (LEP), between 1988 and 1990 there were between 50 and 100 isolates, in 1991 over 200 and in the first 10 months of 1995 in excess of 50% of isolates of DT 104 were of this R-type (LEP, unpublished results - provisional figures only). Molecular studies have demonstrated that in DT 104 all these resistance genes are inserted into the chromosome (Threlfall *et al.*, 1995), which is a rare phenomenon for *S. typhimurium*. Although the organism does not appear more invasive than other types of *Salmonella*, severe illness has been reported in a high proportion of cases.

Many patients have required hospital admission and several deaths have been reported (Wall *et al.*, 1994). Unlike *S. enteritidis* phage type (PT) 4, which is almost entirely associated with poultry and poultry products, multi-resistant *S. typhimurium* DT 104 appears to be widely distributed in food animals. The organism has been isolated from a wide range of food products and in 1993 a Public Health Laboratory Service (PHLS) case-control study showed an association between illness and the consumption of several food items including pork sausages and chickens (Wall *et al.*, 1994), which reinforced the hypothesis that multiple food vehicles are involved in the transmission of this organism to humans. A more recent PHLS study has demonstrated a high incidence of multi-resistant DT 104 in fresh raw sausages purchased from a range of retail outlets in the UK (Nichols and de Louvois, 1995). Although commonly associated with bovine animals, in which it gives rise to high mortality in all ages of livestock (Anonymous, 1993), multiresistant DT 104 has also been isolated from sheep, pigs, chickens and turkeys (LEP, unpublished data). The strain has also been isolated from farm workers who have had direct contact with infected animals, indicating that there is an occupational risk for such people (Wall *et al.*, 1995a), and from domestic pets (Wall *et al*, 1995b). Furthermore, since 1992 there has been an increasing number of strains which, in addition to possessing chromosomally-encoded resistance to ACSSuT, have also acquired plasmid-mediated resistance to sulphonamides and trimethoprim (R-type ACSSuTTm) (Threlfall, *et al.,* 1996). Indeed, in 1994 12% of isolates of DT 104 were of R-type ACSSuTTm and in the first ten months of 1995 over 20% of multiresistant isolates of DT 104 were also resistant to trimethoprim. Although antibiotic therapy is not advocated for most gastrointestinal infections caused by non-typhoidal *Salmonella*, should therapy be required trimethoprim is a useful agent for the first-line treatment of infection. A particularly alarming development in 1995 has been a substantial increase in multiresistant DT 104 with additional chromosomally-encoded resistance to ciprofloxacin (>5% of isolates in the first 10 months of 1995), which is now the drug of choice for the treatment of invasive salmonellosis. It is noteworthy that although resistance to ACSSuT and to ciprofloxacin is chromosomally-integrated, resistance to trimethoprim and to gentamicin/apramycin is plasmid-mediated.

A further recent development of particular concern to public health is an increased incidence, in certain serotypes of *Salmonella*, of strains resistant to ciprofloxacin. In particular, in 1994 40% of isolates of the poultry-associated serotype *S. hadar* and 5% of *S. virchow*, also poultry-associated, were resistant to ciprofloxacin (Frost *et al.*, 1994). Ciprofloxacin is now the therapeutic agent of choice for invasive salmonellosis and the increased incidence of resistance to this antibiotic in commonly-isolated serotypes is alarming. Of particular concern is the increased incidence of strains of *S. virchow* with resistance to this antimicrobial, as this serotype is known to have an enhanced invasive potential for humans

(Threlfall *et al.*, 1990). It has been suggested that the recent licensing of enrofloxacin for veterinary use in the UK may encourage the persistence and spread of ciprofloxacin-resistant salmonellas in food animals (Frost *et al.*, 1995).

## Discussion and conclusions

As antibiotics are not recommended for the treatment of mild to moderate salmonella-induced enteritis in humans it can be argued that drug resistance in food animal-associated, non-typhoidal salmonellas is of little consequence to human public health. However, it is known that overall, between 0.5% and 2% of salmonellas isolated from humans are from blood culture with a higher incidence in certain serotypes, eg *S. virchow* (Threlfall *et al.*, 1992). Antibiotics are also used for the treatment of salmonellosis in immunocompromised patients and sometimes for treating particularly vulnerable patients. In such cases, treatment with an appropriate antibiotic is often essential and can be life-saving. A further facet of antibiotic resistance in zoonotic salmonellas is the undoubted effect that such resistance can have on the dissemination and persistence of multi-resistant strains in such animals, thereby contributing to the increased incidence of such strains in humans. For example, in the multi-resistant *S. typhimurium* DT 204/193/204c outbreak in the late 1970s and early 1980s, there was strong evidence to suggest that the prophylactic use of antibiotics helped provide an environment which enabled the multi-resistant strains to propagate and become widely disseminated in calf herds. It is worth noting that such strains were resistant to almost all of the commonly-used antibiotics and were therefore untreatable, resulting in an extremely high rate of mortality in affected animals (Anonymous, 1985). This in itself must be of concern to farmers and veterinarians.

In conclusion, it would appear that although antibiotics, when used judiciously, may be of considerable benefit to animal health, the injudicious use of such agents in food animals has encouraged the spread and persistence of multi-resistant salmonellas in such animals. This has undoubtably contributed to the overall occurence of such strains in humans and has also resulted in the appearance and spread of strains resistant to antibiotics which are used for the treatment of invasive disease in humans.

## References

Anderson, E.S. (1968) Drug resistance in *Salmonella typhimurium* and its implications. *British Medical Journal*, **5614,** 333-339.

Anonymous (1969) Report of the joint committee on the use of antibiotics in animal husbandry and veterinary medicine. London: HMSO, 1969.

Anonymous (1985) Animal Salmonellosis. 1984 Annual Summaries. Ministry of Agriculture, Fisheries and Food, Welsh Office Agriculture Department, Department of Agriculture and Fisheries for Scotland.

Anonymous (1992) Report of the expert group on animal feedingstuffs. London: HMSO.

Anonymous (1993) Salmonella in Animal and Poultry Production 1992. Ministry of   Agriculture, Fisheries and Food, Welsh Office Agriculture Department, Department of Agriculture and Fisheries for Scotland.

Frost, J.A., Threlfall, E.J. and Rowe, B. (1995) Antibiotic resistance in salmonellas from humans in England and Wales: the situation in 1994. *PHLS Microbiology Digest,* **12,** 131-133.

Nichols, G.L. and de Louvois, J. (1995) The microbiological quality of raw sausages sold   in the UK. *PHLS Microbiology Digest* **12,** 236-242.

Rowe, B. and Threlfall, E.J. (1984) Drug resistance in gram negative aerobic bacilli. *British Medical Bulletin,* **40,** 68-76.

Threlfall, E.J., Ward, L.R. and Rowe, B. (1978a) Epidemic spread of a chloramphenicol-resistant strain of *Salmonella typhimurium* phage type 204 in bovine animals in Britain. *Veterinary Record,* **103,** 438-440.

Threlfall, E.J., Ward, L.R. and Rowe, B. (1978b) The spread of multiresistant strains of   *Salmonella typhimurium* phage types 204 and 193 in Britain. *British Medical Journal* **6143,** 997-998.

Threlfall, E.J., Ward, L.R., Ashley, A.S. and  Rowe, B. (1980) Plasmid-encoded trimethoprim resistance in multiresistant epidemic *Salmonella typhimurium* phage types 204 and 193 in Britain. *British Medical Journal,* **6225,** 1210-1211.

Threlfall, E.J., Rowe, B., Ferguson, J.L. and Ward, L.R. (1985) Increasing resistance to gentamicin and related amino-glycosides in *Salmonella typhimurium* phage type 204c in England, Wales and Scotland. *Veterinary Record,* **117,** 355-357.

Threlfall, E.J., Rowe, B., Ferguson, J.L. and Ward, L.R. (1986) Characterization of   plasmids conferring resistance to gentamicin and apramycin in strains of *Salmonella typhimurium* phage type 204c isolated in Britain. *Journal of Hygiene,* **97,** 419-426.

Threlfall, E.J., Hall, M.L.M. and Rowe, B. (1992) Salmonella bacteraemia in England and Wales, 1981 - 1990. *Journal of Clinical Pathology,* **45,** 34-36.

Threlfall, E.J., Rowe, B. and Ward, L.R. (1992) Recent changes in the occurrence of antibiotic resistance in *Salmonella* isolated in England and Wales. *PHLS Microbiology Digest,* **9,** 69-71.

Threlfall, E.J., Rowe, B. & Ward, L.R. (1993) A comparison of multiple drug resistance in salmonellas from humans and food animals in England and Wales, 1981 and 1990. *Epidemiology and Infection* **111**: 189-197.

Threlfall, E.J., Frost, J.A., Ward, L.R. and Rowe, B. (1994) Epidemic in cattle of *S typhimurium* DT 104 with chromosomally-integrated multiple drug resistance. *Veterinary Record,* **134,** 577.

Threlfall, E.J., Frost, J.A., Ward, L.R. and Rowe, B. (1996) Increasing spectrum of resistance in multiresistant *Salmonella typhimurium. Lancet,* **347,** 1053–1054.

Wall, P.G., Morgan, D., Lamden, K., Ryan, M., Griffin, M., Threlfall, E.J., Ward, L.R. and Rowe, B. (1994) A case-control study of infection with an epidemic strain of multiresistant *Salmonella typhimurium* DT 104 in England and Wales. *Communicable Disease Report,* **4,** R130-R135.

Wall, P.G., Morgan, D., Lamden, K., Griffin, M. & Threlfall, E.J., Ward, L.R. and Rowe, B. (1995a) Transmission of multi-resistant *Salmonella typhimurium* from cattle to man. *Veterinary Record,* **136,** 591-592.

Wall, P.G., Davis, S., Threlfall, E.J. & Ward, L.R. (1995b) Chronic carriage of multidrug resistant *Salmonella typhimurium* in a cat. *Journal of Small Animal Practice,* **36,** 279-281.

Ward, L.R., Threlfall, E.J. and Rowe, B. (1990) Multiple drug resistance in salmonellas isolated from humans in England and Wales: a comparison of 1981 with 1988. *Journal of Clinical Pathology,* **43,** 563-566.

**II**

**Poultry Nutrition**

# 4

# AMINO ACID PROFILES FOR POULTRY

M. PEISKER

*ADM Bioproducts, Auguste-Viktoria-Strasse 16, D - 65185 Wiesbaden, Germany*

## Introduction

The availability of crystalline amino acids has rekindled interest in examining the amino acid requirements for poultry. On a commercial level lysine, threonine, tryptophan and methionine are economically viable although all essential amino acids can be produced as pure substances. The use of crystalline amino acids not only enables the nutritionist to comply better with constraints in linear programming (least cost formulation) and lowering costs but also might contribute significantly to reducing nitrogen output from livestock production.

As adressed in the EU - 1991 council directive concerning the protection of waters against pollution caused by nitrates from agricultural sources (nitrate directive - 91/676 /EEC, annex 3) the amount of manure being spread per hectare may not contain more than 170 kg N annually. Environmental considerations are gaining importance first of all in areas with dense livestock, where the area for spreading manure is limited. Better knowledge of amino acid requirements therefore allows the reduction of dietary protein contents without compromising performance and, at the same time, reducing metabolic stress for the animals.

Specific amino acid requirements depend on a multitude of factors, e.g. sex, strain, crude protein level, ME - level, maximizing feed efficiency, maximizing live weight gain, heat stress, voluntary feed intake and body composition.

### SEX

Usually the amino acid requirement for males is greater than for females. This was proved for arginine and tryptophan (Hunchar and Thomas, 1976), threonine and methionine (Thomas *et al.*, 1986,1987) and lysine (Han and Baker, 1993). Males have higher genetic potential for lean tissue growth, in other words, the maximum N-retention capacity is higher. This is determined by the hormonal differences between the sexes.

## STRAIN

Genetic differences in growth rate are well documented (Moran *et al.*, 1990; Han and Baker, 1991). Fast growth combined with high protein tissue accretion tends to higher amino acid requirements. The key to the prediction as to whether different broiler strains have different amino acid requirements lies in body composition. If a given strain has more protein and less fat in its weight gain than some other strain, then the leaner bird will have higher amino acid requirement.

## DIETARY FACTORS

Requirements for essential amino acids are directly related to dietary energy content. Expressing the level of essential amino acids as % of metabolizable energy (ME) or g digestible amino acid per MJ ME (RPAN 1993) should describe the relationship between dietary energy and amino acid level. Regardless of sex, the lysine requirement for maximum feed efficiency is higher than for maximum live weight gain (Han and Baker, 1993).

## FEED INTAKE / HEAT STRESS

Voluntary feed intake does not necessary affect the lysine requirement. Fast growing strains eat more than slow growing strains (Han and Baker, 1991) but the lysine requirement as a proportion of the diet is identical. Heat-stress can affect voluntary feed intake. Heat stressed birds (37°C) ate less feed than birds housed in a more comfortable environment (24°C) (Han and Baker, 1993a; Maurice, 1995). Heat stress reduced weight gain of males and females by about 22% and increased the lysine requirement for female but not for male chicks.

## BODY COMPOSITION

The protein to fat ratio in the dry matter of chick carcasses can be affected by the dietary lysine level (Liebert, 1995). The lysine level required to maximise breast yield is similar to that required to maximize feed efficiency (Han and Baker, 1994a). Body composition is highly correlated to strain, thus determining the amino acid requirement as such.

# The concept of ideal protein

The various factors which might influence the amino acid requirement need completely different tables of data than those currently in use. The concept of ideal protein was established to account for the multitude of dietary and environmental factors.

Pigs, poultry and other non-ruminant animals have a nutritional requirement not for intact protein, but rather for the essential amino acids that are contained in their dietary crude protein. Two questions have to be answered:

1)   What is the dietary profile or balance of amino acids for a certain species or age?
2)   What is the requirement for one of the essential amino acids within this profile under the different conditions mentioned above?

When question 1 can be answered the ideal protein balance is established. Ideal protein is defined as the perfect profile or balance in terms of dietary concentrations among the essential amino acids. Theoretically, optimal performance should be obtained with the diet that meets all amino acid requirements with no excesses or deficiencies.

Ideal protein also provides the base for a deductive approach to estimate requirements. Given an established ideal profile of amino acids it is a simple matter to calculate the quantitative needs for the remaining nine essential amino acids if the lysine requirement is known.

Lysine was chosen as a reference for ideal protein for several reasons (Baker and Han, 1994a):

1)   Following the sulphur-containing amino acids (SAA), lysine is the second limiting amino acid in broiler diets.
2)   Analysis of lysine in feedstuffs is straightforward.
3)   Lysine has only one function in the body, i.e. protein accretion, thus it is not influenced by the relative proportions of maintenance and growth.
4)   There is a large body of information on the lysine requirement of birds under a variety of dietary, environmental and body composition circumstances.

## Profiles of ideal protein for broiler chicks

Most of the studies establishing ideal ratios of essential amino acids to lysine have been undertaken with chicks between hatching and 21 days post-hatching.

In the period from 21 to 42 days ideal ratios to lysine for some amino acids like SAA, threonine and tryptophan have to be higher due to changing maintenance requirements.

In young birds the maintenance requirement as a percentage of total requirement is very small but increases as birds advance in age and weight. Thus different profiles are established representing the ideal ratio for early and late growth (Table 4.1). These figures are based mainly on work of Baker and Han, University of Illinois, thus the name Illinois Ideal Chick Protein (IICP).

**Table 4.1** IDEAL PROTEIN PROFILE AND DIGESTIBLE AMINO ACID REQUIREMENTS OF BROILER CHICKENS (IICP based on Baker & Han, 1994a/b)

| | 0 ... 21 days | | | 22...42 days | | |
|---|---|---|---|---|---|---|
| | | % of diet | | | % of diet | |
| | Profile | Male | Female | Profile | Male | Female |
| Lysine | 100 | 1.12 | 1.02 | 100 | 0.89 | 0.84 |
| M+C | 72 | 0.81 | 0.74 | 75 | 0.67 | 0.63 |
| Methionine | 36 | 0.405 | 0.37 | 37 | 0.33 | 0.31 |
| Arginine | 105 | 1.18 | 1.07 | 105 | 0.93 | 0.88 |
| Valine | 77 | 0.86 | 0.79 | 77 | 0.69 | 0.65 |
| Threonine | 67 | 0.75 | 0.68 | 70 | 0.62 | 0.59 |
| Tryptophan | 16 | 0.18 | 0.16 | 17 | 0.15 | 0.14 |
| Isoleucine | 67 | 0.75 | 0.68 | 67 | 0.6 | 0.56 |
| Histidine | 32 | 0.36 | 0.33 | 32 | 0.28 | 0.27 |
| PHE + TYR | 105 | 1.18 | 1.07 | 105 | 0.93 | 0.88 |
| Leucine | 109 | 1.22 | 1.11 | 109 | 0.97 | 0.92 |

Taking different tables of requirements, e.g. NRC,1994 (Table 4.2) or Rhone-Poulenc -Animal Nutrition, 1993 (Table 4.3), the recommended amino acid profile turns out to be different. Slight alterations occur further when working with digestible instead of total amino acids (Table 4.3).

In Table 4.4 the amino acid profiles of the IICP, RPAN and NRC are compared for methionine + cystine, methionine, threonine and tryptophan. The most notable difference is for methionine between the IICP and RPAN/NRC. For threonine IICP and RPAN are close, whereas NRC is higher than RPAN for total threonine. The difference in tryptophan between IICP and RPAN is also rather large (19%).

**Table 4.2** AMINO ACID PROFILE AND REQUIREMENTS OF BROILER CHICKENS (NRC 1994)

|  | *0 ... 21 days* | | *22 ... 42 days* | |
|  | *Profile* | *% of diet* | *Profile* | *% of diet* |
|---|---|---|---|---|
| Lysine | 100 | 1.10 | 100 | 1.00 |
| M+C | 82 | 0.90 | 72 | 0.72 |
| Methionine | 46 | 0.50 | 38 | 0.38 |
| Arginine | 113 | 1.25 | 110 | 1.10 |
| Valine | 82 | 0.90 | 82 | 0.82 |
| Threonine | 73 | 0.80 | 74 | 0.74 |
| Tryptophan | 18 | 0.20 | 18 | 0.18 |
| Isoleucine | 73 | 0.80 | 73 | 0.73 |
| Histidine | 32 | 0.35 | 32 | 0.32 |
| PHE + TYR | 122 | 1.34 | 122 | 1.22 |
| Leucine | 109 | 1.20 | 109 | 1.09 |

**Table 4.3** AMINO ACID PROFILE AND REQUIREMENTS OF BROILER CHICKENS (Rhone-Poulenc Nutrition Guide 1993)

|  | 0 ... 21 days | | | | 22 ... 42 days | | | |
|  | Total amino acids | | (3100 kcal/kg M.E.) Digestible amino acids | | Total amino acids | | (3200 kcal/kg M.E.) Digestible amino acids | |
|  | *Profile* | *% of diet* | *Profile* | *% of diet* | *Profile* | *% of diet* | *Profile* | *% of diet* |
|---|---|---|---|---|---|---|---|---|
| Lysine | 100 | 1.18 | 100 | 1.00 | 100 | 1.05 | 100 | 0.89 |
| M+C | 77 | 0.91 | 79 | 0.79 | 79 | 0.83 | 81 | 0.72 |
| Methionine | 47 | 0.55 | 51 | 0.51 | 44 | 0.46 | 48 | 0.43 |
| Arginine | 110 | 1.30 | 117 | 1.17 | 103 | 1.08 | 108 | 0.97 |
| Valine | 83 | 0.98 | 84 | 0.84 | 85 | 0.89 | 85 | 0.77 |
| Threonine | 64 | 0.76 | 65 | 0.65 | 67 | 0.70 | 67 | 0.60 |
| Tryptophan | 19 | 0.22 | 19 | 0.19 | 19 | 0.20 | 19 | 0.17 |
| Isoleucine | 75 | 0.89 | 78 | 0.78 | 72 | 0.76 | 75 | 0.67 |
| Leucine | 140 | 1.65 | 150 | 1.50 | 134 | 1.41 | 144 | 1.28 |

**Table 4.4** COMPARISON OF AMINO ACID PROFILES FOR CHICKENS (0-21 DAYS)

| | Baker & Han 1994 (digestible amino acids) | RPAN (digestible amino acids) | RPAN (total amino acids) | NRC (total amino acids) |
|---|---|---|---|---|
| Lysine | 100 | 100 | 100 | 100 |
| Met + Cys | 72 | 79 | 77 | 82 |
| Methionine | 36 | 51 | 47 | 46 |
| Threonine | 67 | 65 | 64 | 73 |
| Tryptophan | 16 | 19 | 19 | 18 |

In order to check the efficacy of the IICP, Baker and Han (1994b) have compared this profile with the NRC 1984 and 1994 profiles, feeding purified corn-soy diets. In the 1994 NRC-profile the estimated lysine requirement was lowered from 12.0g/kg (1984) to 11.0g/kg of the diet. Estimated requirements for arginine, leucine, cystine, tryptophan and glycine + serine were lowered as well, whereas that for valine was increased. These changes were beneficial. Chicks performed markedly better when fed with the NRC 1994 profile than when fed the NRC1984 profile.

IICP uses lower ratios compared with NRC (1994). In the comparison assay diets contained 9.0g/kg lysine, which was considered slightly deficient for birds fed the purified diet. There were no significant differences in any of the response criteria (weight gain, feed intake, gain to feed), proving that the lower ratios of IICP are adequate under the experimental circumstances.

IICP is based upon digestible whereas NRC is based on total amino acid requirements. This difference should have minimal effect on the ratio comparisons for SAA and threonine, because the estimated true digestibilities of these amino acids are about the same in corn-soybean meal diets, i.e. 88% (Parsons, 1991). For the other amino acids, if using digestible rather than total requirement data for NRC 1994, it does not lower the ratios shown in Table 4.2.

Practical consequences from this work could be summarised as follows:

- under conditions of slightly limiting lysine content the NRC (1994) and the IICP-ratio are of similar efficiency.

- the NRC (1994) lysine requirement is considerably below the Illinois requirement for maximum feed efficiency (Han and Baker,1991, 1993).

- if the lysine requirement for mixed sexes is set for maximum feed efficiency (12.2g/kg diet) then the ratio of NRC (1994) becomes much closer to the IICP (table 4.5).

**Table 4.5** IICP - AND NRC - RATIO WITH LYSINE REQUIREMENT SET FOR MAXIMUM FEED EFFICIENCY

|  | *Baker & Han 1994 (1.07% dig. Lys)* | *Modified NRC (1.22% total Lys)* |
|---|---|---|
| Lysine | 100 | 100 |
| Met + Cys | 72 | 74 |
| Methionine | 36 | 41 |
| Arginine | 105 | 102 |
| Valine | 77 | 74 |
| Threonine | 67 | 66 |
| Tryptophan | 16 | 16 |
| Isoleucine | 67 | 66 |
| Histidine | 32 | 29 |
| PHE + TYR | 105 | 110 |
| Leucine | 109 | 98 |

## Amino acid requirements for broiler chicken

Austic (1994) gave an update upon the amino acid requirements and ratios for broiler chickens (Table 4.6). Lysine is set at 13.0g/kg of the diet, higher than the IICP-recommendation for males.

**Table 4.6** AMINO ACID REQUIREMENTS OF MALE BROILERS TO THREE WEEKS (Austic, 1994)

|  | *% of diet* | *% of protein* | *Amino acid profile* |
|---|---|---|---|
| Lysine | 1.30 | 5.6 | 100 |
| Methionine | 0.50 | 2.2 | 38 |
| Arginine | 1.25 | 5.4 | 96 |
| Valine | 0.90 | 3.9 | 69 |
| Threonine | 0.80 | 3.5 | 62 |
| Tryptophan | 0.24 | 1.1 | 18 |
| Isoleucine | 0.84 | 3.7 | 65 |
| Histidine | 0.32 | 1.4 | 24 |
| Leucine | 1.20 | 5.2 | 92 |

(diet with 23% crude protein/3200 kcal/kg M.E.)

Arginine is set at 12.5g/kg. Excess dietary lysine could increase the arginine requirement. The ratio lysine to arginine should be approximately 1.25 : 1 without adverse effects on performance. With 12.5g arginine/kg diet lysine levels up to 15.0g/kg are tolerable.

Threonine in the range of NRC (1994) with 8.0g/kg in a 230g crude protein/kg diet seems adequate for chicks. With higher crude protein levels the percentage in crude protein is lowered. A value of 35.0g/kg protein appears appropriate for chicks receiving 230gCP/kg but the value may increase slightly as the dietary CP level falls to 200g/kg.

Tryptophan is fixed at 11.0g/kg CP within a crude protein range from 160 - 230g/kg. This corresponds to 2.4g tryptophan/kg diet which is higher than most recommendations found in the literature.

## Amino acids profiles for turkeys

Much less information about the ideal protein profile is available for turkeys. Potter (1989) indicated the limiting order of amino acids for turkeys. Next to methionine and lysine were threonine, valine and isoleucine. Waibel *et al.* (1995) performed trials minimizing the CP content but supplementing methionine and lysine. They used the NRC (1984) data for lysine requirements. It was possible to reduce the protein content in corn-soybean-meal diets to 90% of NRC (1984) if methionine was supplemented throughout and lysine was supplemented after 12 weeks of age each to 100% of NRC-requirement. This was also proved by Spencer (1984) and Sell (1993).

The other essential amino acids seem to be present in adequate quantities at this protein level. Further reduction of the protein level to 80% of NRC resulted in threonine deficiency. With further reduction to 70% of normal no response from threonine addition was observed (Liu *et al.*, 1987).

The deficiencies of other amino acids were probably too severe in this case.

The lack of empirical data on requirements beyond lysine and SAA is probably responsible for the use of minimum crude protein specifications in practical turkey formulations.

In further trials the IICP was used to decide which and how much of each amino acid to supplement. Though the amino acid profile is dependent on age it should be necessary to establish several profiles for turkeys.

Waibel *et al.* (1994) could not substantiate the concern that the NRC (1984) amino acid requirement levels are inadequate for modern turkey strains. However they recommend that the basis be amended to the real attained live weight. NRC (1984) values based on age might underestimate the requirements when the marketing weight is reached earlier and thus suggest higher dietary levels.

Table 4.7 shows the requirements for methionine, cystine, lysine, threonine and valine. The estimates for SAA seem to be well established. Waibel *et al.* (1994) found a somewhat lower estimate for lysine than recommended by NRC (1994). For threonine the NRC requirement is higher up to 9 weeks of age and very close thereafter. The other essential amino acid levels are higher than NRC-values.

**Table 4.7** REQUIREMENTS OF SAA, LYSINE, THREONINE AND VALINE FOR GROWING TURKEYS

| | | | | | | |
|---|---|---|---|---|---|---|
| *(a)* Weeks | *0 ... 4* | *4 ... 8* | *8 ... 12* | | *12 ... 16* | *16 ... 20* |
| *(b)* | *0 ... 3* | *3... 6* | *6 ...9* | *9 ... 12* | *12 ... 15* | *15 ... 18* |
| Meth + Cys | | | | | | |
| (a) | 1.05 | 0.95 | 0.80 | | 0.65 | 0.55 |
| (b) | 1.05 | 0.95 | 0.80 | | 0.65 | 0.55 |
| Lysine | | | | | | |
| (a) | 1.60 | 1.50 | 1.30 | | 1.00 | 0.80 |
| (b) | 1.44 | 1.39 | 1.28 | 1.08 | 0.93 | 0.79 |
| Threonine | | | | | | |
| (a) | 1.00 | 0.95 | 0.80 | | 0.75 | 0.60 |
| (b) | 0.94 | 0.92 | 0.86 | 0.79 | 0.67 | 0.61 |
| Valine | | | | | | |
| (a) | 1.20 | 1.20 | 0.90 | | 0.80 | 0.70 |
| (b) | 1.32 | 1.28 | 1.21 | 1.07 | 0.93 | 0.85 |

a) NRC (1994)    b) Waibel *et al.* (1994)

Table 4.8 presents a calculation for the ideal protein ratio for turkeys (0 - 3 weeks), based on the work of Waibel *et al.* (1994) and compares this with the IICP of Baker and Han (1994) and the NRC (1994). The SAA to lysine and the threonine to lysine are very similar for the two species between the IICP and the NRC (1994). All other amino acids have higher ratios to lysine for turkeys than for the broiler chicken.

The NRC (1994) profile for turkeys from 0 - 3 weeks is, with the exception of methionine + cystine, closer to the IICP ratios. Waibel *et al.* (1994) set the lysine requirement 10% below the NRC (1994) recommendation, which thus explains the different ratios, because all other essential amino acids are at adequate levels in the NRC norms.

**Table 4.8** IDEAL PROTEIN RATIOS FOR TURKEYS (0-3 WEEKS) CALCULATED FROM WAIBEL *et al* (1994) AND COMPARED WITH BAKER & HAN 1994 (IICP) AND NRC (1994)

|  | *Waibel et al (1994)* | *IICP (1994)* | *NRC (1994)* |
|---|---|---|---|
| Lysine | 100 | 100 | 100 |
| Met + Cys | 73 | 72 | 66 |
| Arginine | 117 | 105 | 100 |
| Valine | 92 | 77 | 75 |
| Threonine | 65 | 67 | 63 |
| Tryptophan | 21 | 16 | 16 |
| Isoleucine | 82 | 67 | 69 |
| Histidine | 43 | 32 | 36 |
| Leucine | 141 | 109 | 119 |

A recommendation for a complete profile for turkeys cannot yet be given. It seems that the IICP profile is suited for SAA and threonine but not for the other essential amino acids. Profiles for older birds are to be handled with caution.

## Limitations on the use of crystalline amino acids

Numerous factors, as shown before, influence the overall lysine requirements of poultry. However, whether or not these factors influence the proportion of lysine requirements which can be met with crystalline L-lysine *vs.* that from natural ingredients is less well understood. Nutritionists are inclined to believe that it is possible to use crystalline L-lysine for balancing formulations deficient in that amino acid. L-lysine -HCl is already widely used for that purpose. Higher levels of methionine and lysine are also added, for example to compensate for lower feed intakes in hot climates or in special feed for bird pigmentation, where considerable amounts of corn gluten meal are used.

However, the gap to be filled in European broiler diets is probably not greater than 4.0g/kg (4kg/t), from which it may be assumed that this amount can be adequately used by the birds. Han and Baker (1993b) have studied excess levels of DL-Methionine and L-Lysine in broiler starter diets and concluded that, for these two amino acids, an average of up to 5.0g/kg has neither negative influence on body weight gain nor on feed conversion ratio (FCR). There was a non-significant reduction in feed intake which in turn improved FCR numerically. More pronounced excesses reduce feed intake significantly which affects body weight gain negatively. In practice, these levels will not be reached. Thus for commercial feed formulation there should be no risk making up the amino acid profile with crystalline or synthetic amino acids.

## Bioavailabilty and amino acid efficiency

Because crystalline amino acids are readily absorbed, it is generally assumed that they are completely available for metabolism. The amino acids in most natural feed proteins are less than completely available. When formulating on a gross amino acid basis there is a difference to be expected in terms of amino acid utilization -- or the efficiency of a single essential amino acid to support protein synthesis at the tissue level -- between protein bound and free synthetic amino acids.

Further the utilization of the same limiting amino acid from different feed stuffs might be different due to effects of antinutritive factors (ANF).

Numerous ANFs are known, e.g. trypsin-inhibitors, tannins, lectins, glucosinolates, gossypol or alkoloids, which mainly affect protein digestibility. Presumably it is a better approach to formulate on the basis of digestible or true digestible amino acids. This is not yet the case in all European countries. The argument against it is the lack of sufficient data and the variation between the findings of different research groups.

An even higher level of predictability for amino acid utilization is possible when losses on a tissue level are further taken into consideration. It is known that, for example, up to 40% even of limiting amino acids are catabolized in metabolism. Furthermore amino acids which are damaged, for example during processing, can be digested and absorbed but not utilized. Thus the efficiency of the limiting amino acid from different feedstuffs with the same limiting amino acid can be different. In such cases the dietary usage of crystalline amino acids is more reliable and predictable and also leads to an overall increase in the efficiency of the dietary protein bound amino acids (Table 4.9).

**Table 4.9** N-BALANCE TRIAL WITH GRADED LEVELS OF L-LYS-HCL (8-21 DAYS)

| Diet | N-intake | N-balance | Relative | Relative |
|---|---|---|---|---|
| | mg/kg met. LW/d | | protein quality | lys-efficiency |
| Basal | 4485 | 1172 | 100 | 100 |
| Basal + 0.5 g Lys | 4927 | 1476 | 116 | 104 |
| Basal + 1.4 g Lys | 4942 | 1605 | 128 | 103 |
| Basal + 2.1 g Lys | 4768 | 1734 | 144 | 108 |
| Basal + 2.8 g Lys | 4828 | 1826 | 153 | 106 |
| Basal + 3.5 g Lys | 5016 | 1989 | 164 | 104 |
| Basal + 4.2 g Lys | 4915 | 1980 | 167 | 102 |

(Liebert *et al.* (1994)

Provided that feeding is on a *ad libitum* basis, no constraints for the efficiency of utilisation of dietary lysine occur. In contrast, the efficiency of the total lysine in the test diets which were supplemented with L-Lysine-HCl was on average 4% better. If this difference is attributed only to the added lysine portion the difference would be 7% (Liebert *et al*. 1994). The reason for the improved lysine efficiency is seen in the better dietary amino acid balance in the lysine added feeds.

An integrative approach for the evaluation of amino acid efficiency of utilisation goes beyond the availability level (true ileal digestible amino acids), because it stresses the effect of the limiting amino acid on N-balance and N-retention.

Mathematical descriptions of the correlation between the concentration of the limiting amino acid and the protein quality are available. The slope of the regression between these two parameters depends only on the utilization level of the limiting amino acid, i.e. the efficiency. For synthetic amino acids the efficiency can be regarded as 100%.

Protein-bound amino acids have lower efficiency. Digestibility coefficients cover that largely but some room for variation still remains, for example non-metabolizable heat damaged amino acids. Comprehensive models can account for this and need more attention in the future.

## Conclusions

1.   Ideal amino acid profiles are independent of the variety of factors influencing amino acid requirements

2.   For growing broilers at different stages fairly well established amino acid profiles are available

3.   The degree of predictability is improved, when using amino acid profiles on the basis of true digestible amino acids

4.   For turkeys amino acid requirement data beyond the pre-starter (0 - 4 weeks) period and for amino acids beyond methionine and lysine is sparse

5.   Turkey performance can be maintained with reductions in dietary protein to approximately 90% of NRC (1994), provided diets are supplemented with methionine and lysine

6.   The use of crystalline or synthetic amino acids within the range of practial nutritional boundaries is not limited

7.    Diet supplementation with crystalline amino acids is improves the efficiency of utilisation of protein bound amino acids

## References

Austic, R.E. (1994) Update on amino acid requirements and ratios for broilers. pp 115–120 in: *Proc. of the Maryland Nutrit. Conference*, College Park MD.

Baker, D.H. and Han, Y. (1994a) Ideal protein for broiler chicks. pp 269–272 in: *Proc. of the Maryland Nutrit. Conference*, College Park,MD.

Baker, D.H. and Han, Y. (1994b) Ideal amino acid profile for chicks during the first three weeks posthatching. *Poultry Science.* **73**:1441–1447

Han, Y. and Baker, D.H. (1991) Lysine requirements of fast and slow growing broiler chicks. *Poultry Science.* **70**:2108–2114

Han, Y. and Baker, D.H. (1993a) Effects of sex, heat stress, body weight and genetic strains on the dietary lysine requirement of broiler chicks. *Poultry Science.* **72**:701–708

Han, Y. and Baker, D.H. (1993b) Effects of excess methionine or lysine for broilers fed a corn-soybean meal diet. *Poultry Science.* **72**:1070–1074

Hunchar, J.G. and Thomas, O.P. (1976) The tryptophan requirement of male and female broilers during the 4 - 7 week period. *Poultry Science.* **55**:379–383

Liebert, F. (1995) Lysinverwertung beim Broiler. *Kraftfutter* **3**: 101-104

Liu, J.K., Waibel, P.E. and Noll, S.L. (1987) Methionine, lysine and threonine supplements for turkeys during 8-12 weeks of age. *Poultry Science.* **66, Suppl.1**:134

Maurice, D. V. (1995) Comparison of expanded versus normal commercial broiler chicken diets. *Test report Clemson Univ.*, Clemson, S.C., unpublished

Moran, E.T., Acar, N. and Bilgili, S.F. (1990) Meat yield of broilers and response to lysine. pp 110–116 in: *Proc. of Arkansas Nutrit. Conf.*,Fayettville,AR.

National Research Council (1984) *Nutrient Requirements of Poultry, 8th revised edtition* , Nat. Academy Press, Washington, DC.

National Research Council (1994) *Nutrient Requirements of Poultry, 9th revised edtition* , Nat. Academy Press, Washington, DC.

Parsons, C.M. (1991) Amino acid digestibility for poultry: Feedstuffs: Evaluation and requirements. pp 1–15 in: *Biokyowa Technical Review No. 1*, Biokyowa Press, St. Louis, MO.

Potter, L.M. (1989) Deficient amino acids in low protein turkey diets. pp 1-4 in: *Proc. of the Minnesota Nutrit. Conference*, MN.

Rhodimet Nutrition Guide (1993) *Feed ingredients formulation in digestible amino acids, 2nd edition 1993* Rhone-Poulenc Animal Nutrition

Sell, J.L. (1993) Influence of metabolizable feeding sequence and dietary protein on performance and selected carcass traits of tom turkeys. *Poultry Science.* **72**:521–534

Spencer, G.K. (1984) Minimum protein requirements of turkeys fed adequate levels of lysine and methionine. *MS-thesis, University of Arkansas,* Fayettville, AR.

Thomas, O.P., Zuckerman, A.I., Farranm M. and Tamplin, C.B. (1986) Updated amino acid requirements of broilers. pp 79–85 in: *Proc. of the Maryland Nutrit. Conf.,*College Park,MD.

Thomas,O.P., Farran, M., Tamplin, C.B. and Zuckerman, A.I. (1987) Broiler starter studies: I. The threonine requirements of male and female broiler chicks. II.The body composition of males fed varying levels of protein and energy. pp 38–42 in: *Proc. of the Maryland Nutrit. Conf.,* College Park, MD.

Waibel, P.E., Carlson, C.W., Liu, J.K., Brannon, J.A. and Noll, S.L. (1995) Replacing protein in corn-soybean turkey diets with methionine and lysine. *Poultry Science.* **74**:1143–1158

**5**

## PHOSPHORUS NUTRITION OF POULTRY

J.D. VAN DER KLIS and H.A.J. VERSTEEGH
*Institute for Animal Science and Health, Department of Nutrition of Pigs and Poultry, Runderweg 2, NL-8219 PK, Lelystad, The Netherlands*

## Introduction

Phosphorus (P) is an essential element involved in energy metabolism of all living organisms and it is necessary for bone development. In poultry nutrition, the dietary phosphorus content should meet the bird's requirement in the respective production phases. Dietary phosphorus originates from plant and animal feedstuffs and from feed phosphates. The major fraction (about two thirds) of plant phosphorus is present as phytate phosphorus. Phytates are salts of phytic acid, an inositol with 1 to 6 phosphate groups giving inositol–1–phosphate (IP–1) to inositol–6–phosphate (IP–6). For decades, this organically bound phosphorus has been considered to be unavailable to single stomached animals. As data for the phytate content in plant feedstuffs were lacking, about 70% of phosphorus in plant feedstuffs was generally considered to be present as phytate–P. However, Table 5.1 illustrates that the phytate–P content is variable between plant feedstuffs. The variability in phytate–P contents within feedstuffs was summarised by Sauveur (1989). The remaining inorganic plant phosphorus, together with phosphorus from non-plant sources are considered to be completely available to the single-stomached animal. This approach is the basis for the following system for estimating phosphorus availability, which is still used in diet formulation:

*Available P = total P - phytate-P*

This system of phosphorus evaluation was adopted in diet formulation in combination with appropriate safety margins, in order to prevent phosphorus deficiencies due to inaccuracies in the evaluation system. Since the mid-eighties the occurrence of environmental pollution problems in regions with intensive

71

livestock production has stimulated the development of a more accurate system of phosphorus evaluation to reduce dietary phosphorus surpluses and thereby improve the efficiency of phosphorus accretion in animal tissues. A greater efficiency would result automatically in a lower level of P excretion in urine and faeces. An accurate system of phosphorus evaluation would enable 1) an improvement in dietary phosphorus availability by appropriate feedstuff selection and 2) a reduction in dietary safety margins. In addition, the efficiency of phosphorus accretion in tissues could also be increased by the improvement of the availability of plant phosphorus by dietary supplementation of microbial phytases. These topics will be dealt with in this paper.

## Evaluation of the phosphorus availability in feedstuffs for poultry diets

### METHOD OF EVALUATION

The phosphorus availability in commonly used feedstuffs are measured in three week old male broilers under standardised conditions:

1.  The test feedstuff is the only dietary source of phosphorus present in the experimental diets. Furthermore, these diets contain feedstuffs with very low phosphorus contents like starch, glucose syrup, soya oil, demineralised whey protein and cellulose. Synthetic amino acids, vitamins and minerals (without P) are added to meet the birds' nutrient requirements.
2.  The experimental diets are standardised at 1.8 g available P (aP)/kg feed (as calculated prior to the experiment). The calcium content is standardised at 5.0 g Ca/kg feed using limestone.
3.  The experimental diets are pelleted (3 mm) without steam addition and fed from 10 days of age onwards. The phosphorus evaluation is based on a three day balance period (from 21 to 24 days of age) in which phosphorus intake (feed) and phosphorus excretion (droppings) are measured quantitatively.
4.  Broilers are housed in metabolism cages (15 birds per 0.45 m² cage). Water and feed are continuously available.

As all basal diet components together contain approximately 0.2 g P/kg feed, which is a maximum of 10% of the total available phosphorus content of the test diet, the measured P availabilities were not corrected for the contribution from the basal diet. This correction was omitted as it hardly affected the outcome of the experiments.

In the case of those feedstuffs which contain only low levels of phosphorus (e.g. tapioca), a slope ratio technique was used in which phosphorus from the test feedstuff was gradually exchanged for monosodium phosphate. The dietary aP content was maintained at 1.8 g/kg. The phosphorus availability of monosodium phosphate was determined separately in the same experiment, employing the method described previously.

The evaluation method was based on marginal dietary levels of available phosphorus (1.8 g aP/kg feed) to minimize phosphorus excretion in the urine. In this situation a three-day phosphorus balance can be considered to be appropriate for the estimation of the P availability in the small intestine. Higher levels of aP in the experimental diets would mask differences between feedstuffs, because an excess of absorbed phosphorus would be excreted in the urine (Günther and Al–Masri, 1988), which results in an underestimation of the actual availability.

## PHOSPHORUS AVAILABILITY IN FEEDSTUFFS

In Tables 5.1 and 5.2, data for P availability in feedstuffs of plant and animal origin and in feed phosphates, as determined at our institute, are given. These data show that in broilers the availability of phosphorus from plant feedstuffs varies between 15 and 72%, from animal feedstuffs between 60 and 75% and finally from feed phosphates between 55 and 92%. Each observation was based on 2 to 5 batches of each named feedstuff presented. Based on the measured availabilities of non–plant phosphorus sources it is obvious that P from these sources is not completely available. In Table 5.1 the sum of the phytate–P and aP contents in plant feedstuffs is calculated as a percentage of total P. As these calculated values exceed the total dietary phosphorus contents in several feedstuffs, they suggest that three week old broilers are capable of utilising phytate–P under the standardised experimental conditions (e.g. for legume seeds and wheat by-products).

Therefore, neither of the assumptions for phosphorus evaluation (as presented in the introduction) seems to be valid. Broilers are probably capable of utilising part of the phytate–P on the one hand while, on the other hand, the availability of inorganic phosphorus is less than 100%.

## DEGRADATION OF PHYTATE PHOSPHORUS

From Table 5.1 it was suggested that phytate phosphorus was degraded by three week old broilers under standardised experimental conditions. The significance of phytate–P degradation is quantified in Table 5.3. Calculations in Table 5.3 are based on an assumed P availability of inorganic phosphorus of 80%, a value which is similar to the phosphorus availability from monocalcium phosphate. Based on these calculations hardly any phytate–P degradation is expected for feedstuffs like rice products and sunflower seed (solvent extracted), while phytate–P degradation exceeded 50% for most legume seeds and wheat.

**Table 5.1** THE PHOSPHORUS AVAILABILITY IN SOME PLANT FEEDSTUFFS, MEASURED IN THREE-WEEK OLD BROILERS.

|  | Total P (g/kg) | Phytate–P | Available P (% of total P) | Sum |
|---|---|---|---|---|
| Beans | 4.9 | 74 | 52 | 126 |
| Lupin | 3.0 | 49 | 72 | 121 |
| Maize | 3.0 | 76 | 29 | 105 |
| Maize gluten feed | 9.0 | 45 | 52 | 97 |
| Maize feed meal | 5.1 | 47 | 50 | 97 |
| Peas | 4.1 | 63 | 41 | 104 |
| Rape Seed | 10.9 | 65 | 33 | 98 |
| Rice bran | 17.2 | 82 | 16 | 98 |
| Soya bean (heat treated) | 5.5 | 64 | 54 | 118 |
| Soya bean meal (solvent extracted) | 7.1 | 61 | 61 | 122 |
| Sunflower seed (solvent extracted) | 11.9 | 65 | 38 | 103 |
| Tapioca | 0.9 | 28 | 66 | 94 |
| Wheat | 3.4 | 74 | 48 | 122 |
| Wheat middlings | 10.8 | 74 | 36 | 110 |

**Table 5.2** THE PHOSPHORUS AVAILABILITY IN SOME ANIMAL FEEDSTUFFS AND FEED PHOSPHATES, MEASURED IN THREE-WEEK OLD BROILERS.

|  | Total P (g/kg) | Available P (% of total P) |
|---|---|---|
| Bone meal | 76 | 59 |
| Fish meal | 22 | 74 |
| Meat meal | 29 | 65 |
| Meat and bone meal | 60 | 66 |
| Calcium sodium phosphate | 180 | 59 |
| Dicalcium phosphate (anhydrous) | 197 | 55 |
| Dicalcium phosphate (hydrous) | 181 | 77 |
| Monocalcium phosphate | 226 | 84 |
| Mono-dicalcium phosphate (hydrous) | 213 | 79 |
| Monosodium phosphate | 224 | 92 |

**Table 5.3** THE AVAILABLE PHOSPHORUS (AP) PHYTATE–P AND NON PHYTATE–P
CONTENTS IN SOME PLANT FEEDSTUFFS AND THE CALCULATED DEGRADATION
OF PHYTATE–P[1] BASED ON EXPERIMENTS WITH THREE-WEEK OLD BROILERS.

| | aP | phytate–P | non phytate–P | phytate–P degradation (%) |
|---|---|---|---|---|
| | | $(g/kg)^2$ | | |
| Beans | 2.5 | 3.6 | 1.3 | 53 |
| Lupin | 2.2 | 1.5 | 1.5 | 80 |
| Maize | 0.9 | 2.3 | 0.7 | 16 |
| Maize gluten feed | 4.7 | 4.0 | 5.0 | 22 |
| Maize feed meal | 2.6 | 2.4 | 2.7 | 20 |
| Peas | 1.7 | 2.6 | 1.5 | 23 |
| Rape Seed | 3.6 | 7.1 | 3.8 | 10 |
| Rice bran | 2.8 | 14.1 | 3.1 | 2 |
| Soya bean (heat treated) | 3.0 | 3.5 | 2.0 | 49 |
| Soya bean meal (solvent extracted) | 4.3 | 4.3 | 2.8 | 61 |
| Sunflower seed (solvent extracted) | 4.5 | 7.7 | 4.2 | 19 |
| Tapioca | 0.6 | 0.2 | 0.7 | 38 |
| Wheat | 1.6 | 2.5 | 0.9 | 46 |
| Wheat middlings | 3.9 | 8.0 | 2.8 | 26 |

[1] The digestibility of phytate–P was calculated as:

$$\frac{\dfrac{aP}{0.80} - \text{non phytate-P}}{\text{phytate-P}} \times 100\%$$

[2] Values derived from Table 5.1

It is well-known that endogenous phytases in wheat can account for phytate-P degradation in the small intestine in wheat containing diets. During the pelleting process at our institute (without steam addition) wheat-phytases are not inactivated. In the legume seed diets, broilers seemed to have some adaptive mechanisms to utilise phytate-P, at least under the standardised experimental conditions.

The ability of broilers to hydrolyse phytate–P is in accordance with data from other sources (e.g. Ballam *et al.*, 1985 and Mohammed *et al.*, 1991). The significance of the phenomenon is dependent on the dietary Ca and aP concentrations. Mohammed et al (1991) showed that phytate–P degradation in broilers up to 4 weeks of age was stimulated in corn/soya bean meal diets by lowering the dietary calcium level (10 g/kg *vs.* 5 g/kg) and/or increasing the content of vitamin D3 (500 *vs.* 50.000 IU/kg). Furthermore, Ballam *et al.* (1985) demonstrated that phytate–P degradation could be initiated by low dietary P levels (1.2 g non-phytate-P/kg feed in combination with 10 g Ca/kg feed).

An experiment was conducted to verify the extent of phytate–P degradation under standardised experimental conditions (as described in the section "method of evaluation") and under more practical aP and Ca levels. The dietary aP content was increased alone (5.0 g Ca/kg and 3.0 g aP/kg) or in combination with Ca (8.3 g Ca/kg and 3.0 g aP/kg). The dietary aP content was adjusted by monocalcium phosphate (MCP); the Ca level was standardised using limestone. Based on Table 5.3, in this experiment two plant feedstuffs were chosen as phytate-P sources: one feedstuff with a high expected phytate–P degradation (soya bean meal); and one with a low value (peas). The results are given in Table 5.4.

**Table 5. 4**  THE DEGRADATION OF INOSITOL-6-PHOSPHATE (IP-6) FROM SOYA BEAN MEAL AND PEAS, AS DETERMINED IN FOUR-WEEK OLD BROILERS.

|  | *Experimental diet* | | *IP-6 degradation (%)* |
| --- | --- | --- | --- |
| *Feedstuff* | *Ca (g/kg)* | *aP (g/kg)* | |
| Soya bean meal | 5.0 | 1.8 | 69[D] |
|  | 5.0 | 3.0 | 58[C] |
|  | 8.3 | 3.0 | 36[B] |
| Peas | 5.0 | 1.8 | 38[B] |
|  | 5.0 | 3.0 | 35[B] |
|  | 8.3 | 3.0 | 28[A] |
|  |  | SSD | *** |
|  |  | SED | 2.5 |

SSD, Statistical Significance of Difference: ***, $P<0.001$, NS: not significant
SED, Standard Error of Difference between two means
[A-D] values with different superscript were significantly different ($P<0.05$)

It is obvious that phytate–P degradation was significant in both feedstuffs. Under standard experimental conditions, about 38% IP-6 was degraded in peas and 69% in soya bean meal. Increasing the dietary aP content caused a small reduction in phytate–P degradation (not significant for peas), while the higher Ca content reduced the phytate–P degradation significantly to 28 and 36% for peas and soya bean meal respectively. It was therefore concluded that the degradation of phytate–P in broilers is significant even at more practical Ca and aP levels, but it should be realised that the values for phytate-P degradation under standardised experimental conditions (Table 5.3) will be maxima in practice.

Phytate–P degradation was also shown in laying hens fed corn/soya bean meal diets with 30 g Ca/kg and 40 g Ca/kg and 3.3 g total P/kg (2.7 g phytate–P/kg). At the lower Ca level 34% phytate was degraded at the lower ileum, which was reduced to 10% at the higher Ca level (Van der Klis *et al.*, 1994). Measurements were performed during shell formation.

## Phosphorus equivalence of phytase

Dietary supplementation with microbial phytase can improve the phytate-P hydrolysis in the small intestine and thereby improve the availability of phytate-P (Simons *et al.*, 1990; Edwards, 1993). The enzyme phytase can therefore be considered as a phosphorus "source". It is usually supplemented into broiler and layer diets in exchange for feed phosphates. Research has been carried out to establish the phosphorus equivalency of phytase in corn/soya bean meal broiler (Simons *et al.*, 1992) and layer diets (Van der Klis *et al.*, 1994). In the broiler experiment monocalcium phosphate-P (MCP-P) was replaced by phytase (Natuphos®, produced by Gist Brocades, The Netherlands). Results from 0-2 weeks of age are shown in Table 5.5. From this table it is clear that the dietary MCP-P content can be lowered from 2.2 to 1.2 g/kg when the diet is supplemented with 500 FTU phytase/kg, although the growth was somewhat reduced compared to the positive control diet (diet 1). A further reduction of the dietary MCP-P content resulted in a lower growth performance of the broilers. The phosphorus deposition (g) was not affected by the experimental treatments.

**Table 5.5** THE EFFECT OF EXCHANGE OF MONOCALCIUM PHOSPHATE BY MICROBIAL PHYTASE ON THE PERFORMANCE AND ON PHOSPHORUS DEPOSITION IN TWO-WEEK OLD BROILERS (SIMONS ET AL, 1992).

| Diet | 1 | 2 | 3 | 4 | 5 | | |
|---|---|---|---|---|---|---|---|
| Ca (g/kg) | 7.5 | 7.5 | 7.5 | 7.5 | 7.5 | | |
| P (g/kg) | 7.3 | 6.3 | 5.8 | 5.3 | 4.8 | | |
| MCP-P (g/kg) | 3.2 | 2.2 | 1.7 | 1.2 | 0.7 | | |
| Phytase (FTU/kg)[1] | 0 | 0 | 250 | 500 | 750 | SED | SSD |
| Growth | 388[A] | 378[B] | 384[AB] | 378[B] | 366[C] | 3.9 | *** |
| FCE | 1.19 | 1.20 | 1.19 | 1.17 | 1.18 | 0.008 | NS |
| P deposition (g) | 12.7 | 11.9 | 11.6 | 11.5 | 11.6 | 0.43 | NS |
| P deposition (%) | 46.4[D] | 50.7[CD] | 53.6[BC] | 57.7[B] | 63.5[A] | 2.42 | *** |
| P excretion | 4.7 | 3.7 | 3.2 | 2.6 | 2.1 | | |
| (g/kg growth) | (126) | (100) | (86) | (70) | (56) | | |

[1] One FTU is the phytase activity that liberates 1 μmol ortho-phosphate from 1.5 mmol of Na-phytate in 1 minute at 30°C and pH 5.5 (Engelen *et al.*, 1994)
SSD, SED, [A-D] see Table 5.4

From 2 weeks of age onwards the dietary calcium level was reduced by 1.0 g/kg and the MCP-P by 0.9 g/kg (as far as possible). Data from 2 to 6 weeks are not shown, as differences between dietary treatments disappeared (average final weight was 2350 g at a feed conversion efficiency of 1.63). Based on this experiment it was concluded that 250 FTU phytase/kg feed is equivalent to 0.5 g MCP-P/kg feed in broilers fed corn/soya diets. This equivalence was valid up to an inclusion level of 500 FTU/kg. As the phosphorus deposition (g) was not affected, a reduction in dietary P level resulted in a significantly improved P deposition (%). When 1.0 g MCP-P/kg feed is exchanged for 500 FTU/kg feed the P excretion/ kg growth was reduced by 30% compared to diet 2 and 56% compared to the positive control diet.

In the laying hen experiments MCP-P and phytase were added separately to the corn/soya bean meal basal diet (total P: 3.2 g/kg; phytate-P: 2.4 g/kg). The dietary Ca contents were standardised at 35 g/kg feed. Results are given in Table 5.6. Measurements were performed during shell calcification, which resulted in a high apparent Ca absorption. The P absorption of the basal diet was low and was significantly increased by either MCP-P addition or phytase supplementation. Thus 0.87 g P/kg feed was absorbed from the supplemented 1.0 g MCP-P/kg feed, while 250 FTU and 500 FTU phytase/kg feed resulted in an improvement in P absorption respectively of 0.70 and 0.92 g. From these data it was calculated that 250 FTU phytase/kg diet was equivalent to 0.8 g MCP-P/kg diet (Figure 5.1). Higher supplementation levels of phytase resulted in a lower equivalence per unit of phytase due to the significant quadratic phytase response in P absorption. This equivalence was similar to a value obtained from a laying hen experiment, which lasted from 18 to 68 weeks of age (Simons and Versteegh, 1993).

**Table 5.6** THE PHOSPHORUS ABSORPTION AT THE LOWER ILEUM OF 24-WEEK OLD LAYING HENS (VAN DER KLIS *et al.*, 1994).

| Diet | MCP-P (g/kg feed) | Phytase (FTU/kg feed) | Ca (%) | P (%) | P (g/kg feed) |
|------|-------------------|-----------------------|--------|-------|---------------|
| | | | *Ileal absorption* | | |
| 1 | 0 | 0 | 72.0 | 26.2[A] | 0.85 |
| 2 | 1.0 | 0 | 74.0 | 40.6[B] | 1.72 |
| 3 | 0 | 250 | 72.6 | 47.7[C] | 1.55 |
| 4 | 0 | 500 | 70.6 | 54.5[D] | 1.77 |
| | | SSD | NS | *** | |
| | | SED | 3.44 | 2.68 | |

SSD, SED, [A-D] see table 5.4

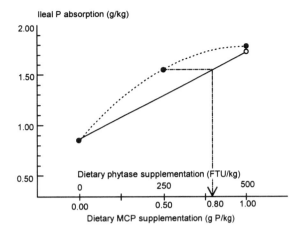

**Figure 5.1** The phosphorus equivalence of microbial phytase in 24 week old laying hens. Dietary supplementation with: ○ Mono calcium phosphate (MCP) ● Phytase

Based on tibia parameters at 68 weeks of age they calculated that 280 FTU/kg feed was equivalent to 1.0 g MCP-P/kg diet. In all experiments carried out so far the equivalence in laying hens (300 FTU = 1.0 g MCP-P or 0.8 g aP) is higher than in growing broilers (250 FTU = 0.5 g MCP-P or 0.4 g aP). The equivalencies with other feed phosphates can be calculated on an availability basis (as shown in Table 5.2).

## Phosphorus requirements

The dietary available phosphorus content should meet the animal's requirement, as on the one hand too low levels cause losses in animal productivity and on the other hand too high levels would result in a reduced efficiency of phosphorus deposition (and higher P contents in droppings as a consequence).

The available phosphorus requirements for growing broilers are presented in Table 5.7. These calculations are based on a factorial approach. The aP requirements were derived for laying hens in a similar manner. It should be realised that these requirements are based on a limited number of carcass P analyses (WPSA, 1985). The aP requirement for laying hens was approximately 2.5 g aP/kg throughout the laying cycle (CVB, 1994).

**Table 5.7**  THE PHOSPHORUS REQUIREMENTS IN BROILERS (MALES AND FEMALES), CALCULATED BY A FACTORIAL APPROACH.

| Period (d)[1] | Live weight (g)[2] | | P in carcass (mg/g LW)[3] | | P main. (mg)[4] | P growth (mg)[5] | Feed intake (g)[2] | aP (g/kg)[6] |
|---|---|---|---|---|---|---|---|---|
| t1 - t2 | t1 | t2 | t1 | t2 | | | | |
| 0 - 10 | 42 | 220 | 3.4 | 4.9 | 21 | 935 | 260 | 3.68 |
| 10 - 30 | 220 | 1235 | 4.9 | 4.8 | 232 | 4850 | 1705 | 2.98 |
| 30 - 40 | 1235 | 1820 | 4.8 | 4.8 | 222 | 2808 | 1290 | 2.35 |

[1]  t1, t2: start and end of period (in days of age)
[2]  Live weight and feed intake data obtained from Ross Breeders
[3]  Source: WPSA (1985)
[4]  P maintenance was based on a calculated endogenous excretion of 12.5 mg P/d in broilers, fed a monosodium phosphate as a P source. Live weight 800g. For a derivation see CVB (1994).
[5]  P deposition in carcass during period t1 - t2
[6]  (P maintenance + P growth)/ feed intake

*These calculations were also carried out for other types of poultry and published by CVB (1994).*


## Optimum dietary calcium/available phosphorus ratio

It has been shown in the previous sections that the availability of phosphorus varies between feedstuffs. The availability of phytate–P can be improved by dietary phytase supplementation. However, the efficiency of dietary phosphorus utilization will be reduced at suboptimal Ca/aP ratios as on the one hand the phytate–P degradation in the gastro-intestinal tract as well as the phosphorus absorption from the small intestine will be reduced at high levels of calcium, while on the other hand absorbed phosphorus will be excreted in the urine if the dietary calcium levels are too low. An experiment was performed to determine the optimum dietary Ca/aP ratio in 2 to 4 week old broilers. Two experimental basal diets were formulated containing 2.5 g aP/kg and 2.0 and 3.0 g phytate-P/kg. The phytate–P containing feedstuffs in the low phytate–P diet were tapioca (41.1%) and soya bean meal (36.5%) and in the high phytate–P diet corn (7.2%), tapioca (25.2%), soya bean meal (34.2%) and sunflower seed meal (10.0%). The Ca/aP ratio was increased stepwise from 1.6 to 2.8 in five increments of 0.3.

The phosphorus absorption (% of intake) and retention (% of absorbed P) is shown in Figure 5.2. At the low dietary calcium level, phosphorus absorption from the small intestine was maximal, but due to the lack of a proper counter ion it was deposited in the body with the lowest efficiency. At increasing dietary calcium levels, phosphorus absorption was reduced and the efficiency of phosphorus retention improved. The efficiency of both calcium and phosphorus retention is maximal at the point of intersection between the Ca/P ratio as absorbed and as retained (Figure 5.3). From this figure an optimum Ca/P ratio (on an availability basis) of 1.25 was calculated for the low phytate–P diet. At the high phytate–P level this optimal value was 1.33. These ratios corresponded to an optimal total calcium/ available phosphorus ratio of 2.2 and 2.3 respectively.

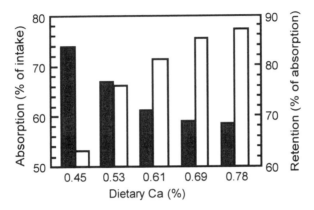

**Figure 5.2** The absorption and retention of phosphorus in four week old broilers, fed a diet containing 2.5 g available P/kg feed and 2.0 g phytate-P/kg feed and variable Ca levels (■ absorption, ☐ retention).

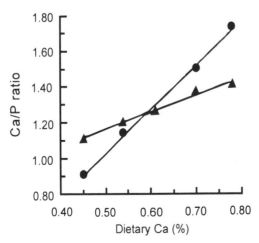

**Figure 5.3** The Ca/P ratio as absorbed and retained in four week old broilers, fed a diet containing 2.5 g available P/kg feed and 2.0 g phytate-P/kg feed (● absorbed, ▲ retained).

## Physico-chemical chyme conditions and mineral absorption

One of the physico-chemical conditions of the chyme (intestinal contents) from the small intestine which has received much attention during the last five years is the viscosity of its fluid phase. There is considerable evidence in broilers that increasing the intestinal viscosities through cereal inclusion in the diet affects the digestion of absorption of nutrients negatively, resulting in a poorer performance than expected from the dietary feedstuff composition. Furthermore, less efficient nutrient utilisation results in higher animal manure production. It was also shown at our institute that mineral absorption might be affected negatively by increasing intestinal viscosities (van der Klis, 1993), where it was shown that the absorption of phosphorus from monocalcium phosphate was reduced in broilers when 1% carboxy methyl cellulose (CMC) was included in a semi-synthetic diet (reduction of 8% at the end of the small intestine in the 1% CMC diet, relative to the diet without CMC). This inverse relationship was also found in wheat-based diets (12% lower phosphorus absorption in high viscosity wheats, compared to low viscosity wheats). The diets in the last experiment contained 50% wheat (Van der Klis, 1993). This potential negative effect should be taken into account if the dietary aP content is calculated to meet the birds' requirements in the respective production phases.

## References

Ballam, G.C., Nelson, T.S. and Kirby, L.K. (1985). Effect of different dietary levels of calcium and phosphorus. *Nutr. Rep. Int.* **32**: 909-913.

CVB (1994). Voorlopig systeem opneembaar fosfor pluimvee. *(Interim System Available Phosphorus for Poultry). Published by the Centraal Veevoeder Bureau, Lelystad, The Netherlands, CVB reeks nr.* **16**.

Edwards, H.M. (1993). Dietary 1,25-dihydroxycholecalciferol supplementation increases natural phytate phosphorus utilization in chickens. *J. Nutr.* **123**: 567-577.

Engelen, A.J., Heeft, F.C. van der., Randsdorp, H.G. and Smit, E.L.C. (1994) Simple and rapid determination of phytase activity. *J. AOAC Intern.* 77:760–764.

Günther, K.D. and Al–Masri, M.R. (1988). Untersuchungen zum Einfluß einer variierten Phosphorus-Versorgung auf den P-Umsatz und die endogene P–Ausscheidung beim wachsenden Geflügel mit Hilfe von [32]P. *J. Anim. Physiol. a. Anim. Nutr.* **59**: 132-142.

Mohammed, A., Gibney, M.J. and Taylor, T.G. (1991). The effects of dietary levels of inorganic phosphorus, calcium and cholecalciferol on the digestibility of phytate–P by the chick. *Br. J. Nutr.* **66**: 251-259.

Sauveur, B. (1989). Phosphore phytique et phytases dans l'alimentation des volailles. *INRA Prod. Anim.* **2**: 343-351.

Simons, P.C.M., Versteegh, H.A.J., Jongbloed, A.W., Kemme, P.A., Slump, P., Bos, K.D., Wolters, M.G.E., Beudeker, F.R. and Verschoor, G.J. (1990). Improvement of phosphorus availability by microbial phytase in broilers and pigs. *Br. J. Nutr.* **64**: 525-540.

Simons, P.C.M., Jongbloed, A.W., Versteegh, H.A.J. and Kemme, P.A. (1992). Improvement of phosphorus availabilities by microbial phytase in poultry and pigs. *In Georgia Nutrition Conference*, Atlanta, November 1992.

Simons, P.C.M. and Versteegh, H.A.J. (1993). Het effect van toevoeging van lage doseringen microbieel fytase aan leghennenvoer op de technische resultaten en de skelet- en eischaalkwaliteit. *Spelderholt Report* **589** (printed in Dutch).

Van der Klis, J.D., Versteegh, H.A.J. and Scheele, C.W. (1994). Practical enzyme use in poultry diets: phytase and NSP enzymes. *In BASF Technical Symposium during the Carolina Poultry Nutrition Conference*, Charlotte, December 1994, pp 113-128.

Van der Klis, J.D. (1993). Physico-chemical chyme conditions and mineral absorption in broilers. *PhD thesis*, Agricultural University, Wageningen, The Netherlands.

WPSA (1985). Mineral requirements for poultry - Mineral recommendations for growing birds. *WPSA Journal* **41**: 252-258.

# III

# Laboratory Analysis

**6**

# THE VALUE OF TRADITIONAL ANALYTICAL METHODS AND NEAR-INFRARED (NIR) SPECTROSCOPY TO THE FEED INDUSTRY

I. MURRAY
*Scottish Agricultural College, 581 King Street, Aberdeen, AB9 1UD, U.K.*

The UK animal feedstuff industry turns over £3 billion per year and employs some 10,000 staff located in 230 feed mills producing 12 million tonnes of compound feed annually of which 41% is destined for ruminants, 34% for poultry and 22% for pigs. It is of strategic importance in supporting both farming and rural communities as primary producers of raw materials, and we, the consumers, who depend on the quality and wholesomeness of the animal products arising from these feeds. The animal feed industry connects the Food and Agriculture sectors of the economy to facilitate the upgrading of raw materials into high quality animal and dairy products for elaboration into human food. Concerns for consumer nutrition, health and safety must begin from the raw ingredients which enter the food chain as animal feed. Food represents our most intimate contact with our environment and has a profound effect on the health, well-being and productivity of the nation. The scale of operations and the volume of material handled, from numerous global sources, create problems in trade, tariffication, traceability, quality control and consumer safety. The industry has undergone transformation into a smaller number of larger operational units with economies of scale, streamlining of production, distribution and cutting of costs. However quality assurance has never been more important to the profitablility of the industry and the consumer confidence on which it depends.

Traditional methods of feed testing, developed originally at the end of the nineteenth century, are no longer acceptable or effective in supervising or regulating this international operation. In testing for the grossly obvious oil, moisture, protein, fibre and ash, there is a danger of allowing more serious deviations in raw materials (e.g. contamination, moulding, overheating) to pass undetected. A radical new approach to qualifying raw materials and finished products is required which can be easily applied at key points in the manufacturing process. One candidate technique has all the hallmarks of providing the solution to industrial need, business management and regulatory surveillance; this is Near Infrared (NIR) Spectroscopy. NIR is a computer based technique which has come of age from the application of multivariate statistical methods to easily acquired reflectance spectra. Often, though incorrectly, described as an expert system, NIR meets the criteria of speed of response, reliability, cost-effectiveness and fitness for purpose. This paper explores the possibilities for further development of NIRS in the feed industry. It updates a previous paper in the series which dealt with the application of NIR in forage analysis (Murray, 1986). In the decade between these papers there have been significant improvements in instrumentation and algorithms (Windham and Flinn, 1992), software and computers (Shenk and Westerhaus, 1993).

Figure 6.1 shows biomass flow in contemporary agricultural ecosystems. Crops grown to produce products for direct human consumption such as cereals, oilseeds, sugar beet etc. yield by-products and waste products which arise at various stages of production and elaboration into human food and drink. A by-product may be defined as a "substance which is unavoidably generated in the elaboration of a more valued product". This economic definition means that more management effort will be concentrated on the quality of the product rather than the by-product. Variability in the product is undesirable so the by-product tends to carry increased burdens of variability. The feed industry has become increasingly reliant upon by-products which are inherently variable. Likewise any crop surplus routed to animal feeding tends to have more variability than that destined for human food. Thus the "straights" contributing to animal feeding are more variable and no good distinction exists been "straight feedingstuffs" "ingredients" and "raw feed materials". The commission of the European Communities has introduced a proposal to re-define these as "feed materials" (CEC 1994).

## Variability

The problem of variability of feed materials used to derive compounded feeds is of very real concern because the uncertainties like errors are cumulative. Brigstocke (1989) demonstrated the variability observed in 37094 tonnes of Maize gluten used by BOCM in 1988 based on 895 samples taken for analysis. (Table 6.1)

Raw materials are not at all homogeneous so that a "book" value or mean value may not pertain to the composition of an individual batch. A book value of 20% protein assumed for a material which is in fact 14% can have a serious effect on the compound feed produced. More frequent analysis is the only way to take account of variability. The major cost of ingredient variability is the cost incurred

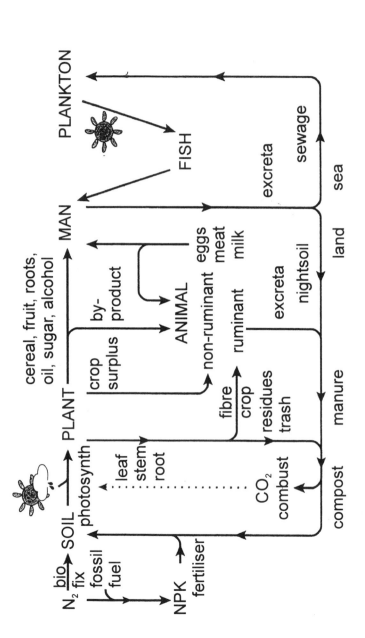

**Figure 6.1** Biomass flow in agricultural ecosystems

Crops grown for human food and drink yield by-products which are unavoidably generated in the elaboration of more highly valued products. By-products are inherently variable but important ingredients recycled as animal feed. As much as 70% of animal production costs are feed costs. Animal performance depends on consistent feeds manufactured from highly variable by-products.

in producing a consistent quality feed from ingredients of variable nutrient composition (Duncan, 1988). Costs are incurred by the feed manufacturer in identifying/analysing for deficiencies, overcoming them through intake segregation and formulation changes and costs incurred by the farmer if performance is lost through failure to meet the specification.

**Table 6.1**   OBSERVED VARIABILITY IN ANALYSIS OF MAIZE GLUTEN BASED ON 895 SAMPLES FROM 37094 TONNES RECEIVED DURING 1988

|          | *Target* | *Mean* | *Range* |
|----------|----------|--------|---------|
| Oil      | 3.0      | 3.2    | 2.0 - 8.4 |
| Protein  | 19.2     | 19.56  | 14.0 - 27.8 |
| Fibre    | 7.2      | 7.11   | 5.5 - 9.1 |
| Ash      | 6.9      | 6.91   | 3.9 - 9.9 |
| Starch   | 20.3     | 21.05  | 12.5 - 23.8 |
| Ca       | 0.35     | 0.18   | 0.09 - 1.52 |
| P        | 0.80     | 0.81   | 0.53 - 1.00 |

Brigstocke (1989)

That variability can be reduced by mixing ingredients is shown in the example in Table 6.2.   If a compound is prepared by mixing equal proportions of 5 ingredients,  the mean value of the mixture is given by $\Sigma m_i x_i$, the sum of the products of the mean nutrient value of each ingredient and the formulation fraction. The variability SD of the mixture is given by

$$\text{SD mix} = \sqrt{(F_1 SD_1)^2 + (F_2 SD_2)^2 + \ldots (F_5 SD_5)^2} = \sqrt{\Sigma(F_i SD_i)^2}$$

where $F_i$ is the fraction of the total nutrient contributed by ingredient i, and $SD_i$ is the standard deviation of the nutrient in ingredient i (Chung and Pfost (1964) cited by Duncan (1988)).

The SD of the mix is inflated by ingredients having a large SD and by the fraction of the total nutrient contributed by an ingredient.  So analytical effort is best directed to those ingredients whose $F \times SD$ products contribute most to the variability of the mix. In the example given, 80% of the variability in the mix derives from ingredient 3.  Increasing the specification of this ingredient will exert the most improvement in the variability of the mix. More frequent sampling and analysis will improve knowledge of the variability of ingredient 3 and allow formulation adjustment to make a more uniform product. Rapid analysis by NIR can make this possible.

Figure 6.2 shows how ingredients having variable nutrient contents can be mixed to form consistent compound feeds with reduced nutrient variability.

**Table 6.2** INGREDIENT VARIABILITY AND ITS EFFECT ON THE VARIABILITY OF A COMPOUND MIX

| Ingredient | Nutrient Mean $m_i$ | Nutrient $SD_i$ | Formulation Fraction $x_i$ | Nutrient Fraction $F_i$ | $(F_i SD_i)^2$ |
|---|---|---|---|---|---|
| 1 | 5 | 2.2 | 0.2 | 1.0/8.0 | 0.0756 |
| 2 | 7 | 1.5 | 0.2 | 1.4/8.0 | 0.0689 |
| 3 | 9 | 4.0 | 0.2 | 1.8/8.0 | 0.81 |
| 4 | 14 | 1.2 | 0.2 | 2.8/8.0 | 0.176 |
| 5 | 5 | 2.5 | 0.2 | 1.0/8.0 | 0.097 |
| | | Sum | 1.0 | 1.0 | 1.228 |

**mean mix = 8.0**      **SD mix** $= \sqrt{1.23} = 1.11$

In this example the SD of the mixture is less than that of any one ingredient.

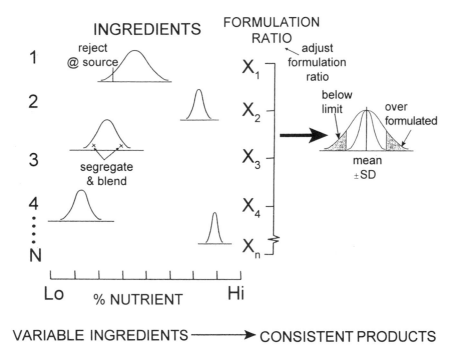

VARIABLE INGREDIENTS ⟶ CONSISTENT PRODUCTS

**Figure 6.2** Formulation of consistent products from variable ingredients
Manufacture of a consistent product having a target mean and minimum SD in key nutrients avoids over or under formulation. Strategies are analysis-enforced specification with rejection, screening or segregation on intake. Batchwise or continuous formulation adjustment compensates for variability. Decision-making depends on fast, accurate analysis on intake and key pionts in manufacture. QC is best directed at most variable ingredients

Strategies for reducing variability are:

1.   setting and enforcing strict specifications on suppliers
2.   screening and cleaning on intake
3.   concentrating QC on most variable ingredients
4.   scheduling high and low batches to arrive simultaneously
5.   quality segregation on intake with controlled blending
6.   batch-wise adjustment of formulation ratios/substitution
7.   continuous adjustment of formulation ratios by closed-loop control.

## Avoiding over formulation

The importance of finished product variability as measured by the standard deviation becomes clear when we consider that a wider SD incurs the double penalty of an increased mass of samples being both under and over formulated. Percentage points of the standardised normal distribution show the values for the mean which must be observed for any given failure rate (Table 6.3).

**Table 6.3**   RELATIONSHIP BETWEEN THE PERCENTAGE OF SAMPLES FAILING TO MEET A MINIMUM NUTRIENT CONTENT AND THE VALUE FOR THE MEAN WHICH MUST BE OBSERVED ASSUMING A NORMAL DISTRIBUTION.

| Failure Rate | Mean must be |
|---|---|
| 20% | 0.84 * SD greater than the minimum |
| 15% | 1.04 * SD greater than the minimum |
| 10% | 1.28 * SD greater than the minimum |
| 5% | 1.65 * SD greater than the minimum |
| 1% | 2.33 * SD greater than the minimum |

e.g. 5% failure rate and minimum 18.0

| SD | 1.65 SD | Observed mean |
|---|---|---|
| 0.2 | 0.33 | 18.33 |
| 0.4 | 0.66 | 18.66 |
| 0.6 | 0.99 | 18.99 |
| 0.8 | 1.32 | 19.32 |
| 1.0 | 1.65 | 19.65 |
| 1.2 | 2.0 | 20.00 |

To ensure 95% of samples are above the minimum, costs 0.66% over formulation at an SD of 0.4; costs 1.32% over formulation at an SD of 0.8; costs 2.0% over formulation at an SD of 1.2.   Clearly over formulation is an expensive means of overcoming variability in ingredients especially if the nutrient has a high unit cost.

Batches of material arriving may be over or under specification such that segregation into 2 bins with subsequent blending may reduce the variability in the final product.   In this case successful segregation does not require an NIR calibration with a high $R^2$. (Table 6.4 from Shenk, 1993).

**Table 6.4** VALUES FOR $R^2$ FOR CALIBRATION AND THE PERCENTAGE OF TIMES THAT A SAMPLE IS CORRECTLY CLASSIFIED INTO 2 OR 3 GROUPS

| $R^2$ | 0.1 | 0.2 | 0.3 | 0.4 | 0.5 | 0.6 | 0.7 | 0.8 | 0.9 |
|---|---|---|---|---|---|---|---|---|---|
| 2 groups (%) | 61.4 | 64.8 | 68.3 | 71.8 | 75.0 | 78.2 | 81.7 | 85.2 | 89.8 |
| 3 groups (%) | 43.2 | 47.1 | 51.0 | 54.9 | 59.0 | 63.1 | 68.4 | 73.6 | 81.4 |

Shenk and Westerhaus (1993)

Duncan (1988) showed that physical separation into two or more storage bins will reduce the variability through blending according to

$S' = S/\sqrt{n}$ where n is the number of bins drawn from
　　　　　S is the standard deviation prior to blending
　　　　　S' is the standard deviation of the blend

An example of segregation of wheat at intake into two bins for wheat samples either above or below 11% in a range from 7.6 to 15.9% crude protein, reduced the SD from 1.12 overall to 0.56 and 0.75 for the low and high bins respectively. (Table 6.5).

**Table 6.5**  SEGREGATION AT INTAKE OF WHEAT BY PROTEIN CONTENT

|  | *All* | *Low <11%* | *High >11%* |
|---|---|---|---|
| Mean | 11.16 | 10.16 | 11.94 |
| SD | 1.12 | 0.56 | 0.75 |
| CV | 10.0 | 5.51 | 6.28 |
| Proportions | 100 | 41 | 59 |
| Range | 7.6 - 15.9 | 7.6 - 11 | 11 - 15.9 |

Ball, (1988)

Segregation however proves costly in terms of supplying storage capacity. A cost which has to be gained either from improved consistency of the blend or from added value arising from better utilisation of nutrients in alternative more competitively priced products. Continuous analysis of ingredient streams entering a production process is being done routinely in the baking, confectionery and chemicals industry to avoid over formulation or segregation. Where the process is continuous, closed-loop feedback control becomes a reality.

## Legal requirements

The starting point for discussion of feed material analysis has to be the legal requirements for declaration (Williams, 1987). Table 6.6 lists the provisions regarding minimum tolerated levels indicated or to be declared where, on inspection, the composition of a feed material is found to depart from the declared composition in a manner such as to reduce its value (CEC 1994). The minimum tolerances are given on a weight basis "as received". In essence this is the "10% of declared" rule with a 12% tolerance for oils and 15% tolerance for crude fibre. The regulations seem to recognise the poorer performance of oil and fibre measurements. The low values in Table 6.6 indicate the level of tolerance which must be achieved by any laboratory method of evaluation whether by wet chemical procedures or by NIRS.

**Table 6.6** MINIMUM TOLERATED VALUES - "THE 10% DECLARED RULE"

Where, on official inspection pursuant to Article 12 of the Directive, the composition of a feed material is found to depart from the declared composition in a manner such as to <u>reduce its value</u>, the following minimum tolerances are permitted:

|  |  | *weight basis "as received"* | |
|---|---|---|---|
|  | *High* | *Medium* | *Low* |
| a) Crude protein | 2   units if ≥20% | 10% of declared 10 - 20% | 1   unit if <10% |
| b) Sugar(s) | 2   units if ≥20% | 10% of declared 5 - 20% | 0.5 unit if  <5% |
| c) Starch, inulin | 3   units if ≥30% | 10% of declared 10 - 30% | 1   unit if <10% |
| d) Crude fat | 1.8 units if ≥15% | 12% of declared 5 - 15% | 0.6 unit if  <5% |
| e) Crude fibre | 2.1 units if ≥14% | 15% of declared 6 - 14% | 0.9 unit if  <6% |
| f) Moisture, crude ash | 1   unit if  ≥10% | 10% of declared 5 - 10% | 0.5 unit if  <5% |
| g) P, Na, CaCO₃, Ca, Mg | 1.5 units if ≥15% | 10% of declared 2 - 15% | 0.2 unit if  <2% |
| h) Ash insol HCl, |  |  |  |
|    Cl as NaCl |  | 10% of declared >3% | 0.3 unit if  <3% |
| i) Carotene, Vit A, |  |  |  |
|    Xanthophyll |  | 30% of declared |  |
| j) Methionine, Lysine, |  |  |  |
|    N bases |  | 20% of declared |  |

Moisture declared if >14.5% (otherwise optional)
Binding agents declared >3%Acid insoluble ash declared if >2.2% DM
Denaturing agents - declare nature and quantity used
CEC (1994)

During the 1980's a number of experiments were conducted at Feed Evaluation Units at the Rowett Research Institute (Wainman *et al.*, 1981) and the ADAS Centre at Stratford-upon-Avon (Alderman, 1985). These feeding trials were conducted to provide ME values for feeds and to devise laboratory methods to predict animal performance from chemical analyses (MAFF, 1993). As a result of these studies the performance of traditional wet chemical analysis has been quite well documented. For example Wainman, Dewey and Boyne (1981) examined 26 ruminant compound feeds to determine their *in vivo* ME values and derive regression equations relating traditional analyses to these. This provided the first opportunity for Murray and Hall (1983) to attempt NIR spectroscopy on compound feeds using only two wavelengths. Hall (1983) attempted to partition the error between the laboratory and the NIR technique (Table 6.7).

**Table 6.7** PREDICTION OF PROPERTIES OF 26 RUMINANT COMPOUND FEEDS FROM NIR SPECTROSCOPY

|  |  | *SEC* | *SE lab* | *SE NIR* |
|---|---|---|---|---|
| Oil | % | 0.37 | 0.17 | 0.329 |
| Protein | % | 0.69 | 0.25 | 0.642 |
| Starch | % | 2.12 | 1.28 | 1.69 |
| ADF | % | 1.64 | 0.60 | 1.53 |
| GE | MJ/kgDM | 0.34 | 0.25 | 0.23 |
| ME in vivo | MJ/kgDM | 0.39 | 0.18 | 0.35 |

SEC = standard error of calibration = $\sqrt{SE^2_{lab} + SE^2_{NIR}}$
Hall (1983)

Hall (1983) showed that as few as two wavelengths could successfully predict the composition of these 26 compound feeds. However the number of samples was inadequate to perform calibrations which were likely to be robust in routine use and since no other analysed materials were then available the study could only show NIR to have potential as an analytical tool. Alderman (1985) reported on a collaborative study of 12 compound feeds analysed in 16 laboratories to establish the variability of statutory analyses (Table 6.8).

**Table 6.8** VARIABILITY IN STATUTORY ANALYSES (12 COMPOUNDS, 16 LABS)

|  | *Crude Fibre* | *Crude Protein* | *Ash* | *Oil* |
|---|---|---|---|---|
| Mean % | 9.05 | 16.94 | 9.28 | 4.14 |
| Repeatability r | 0.44 | 0.49 | 0.3 | 0.31 |
| Reproducibility R | 1.45 | 1.41 | 0.74 | 0.65 |

Alderman (1985)

Estimates of precision of traditional wet chemical analysis become larger among groups of laboratories working under routine conditions. Alderman (1985) showed values for the reproducibility R as a coefficient of variation of 4.2 for crude fibre, 1.6 for crude protein, 2.1 for ash and 4.9 for crude lipid. These values pertain to the ADAS and UKASTA laboratories in 1984 (UKASTA, ADAS and COSAC, 1985). In NIR analysis, standard errors of estimate (SEC, SEP, SECV) are used as measures of precision (See Appendix 1).

## Traditional analysis

Traditional analyses for composition using wet chemical methods began at the end of the 19th century when the Weende Proximate Analysis scheme was developed by Henneberg & Stohmann (1864) and Kjeldahl (1883) and subjected to much modification over the 20th century. Some 60 official methods of analysis are detailed in The Feeding Stuffs Regulations and Amendments (Statutory Instruments, 1982,84,85). While these are important reference methods for legal purposes, they are viewed as being inappropriate, inconvenient, uneconomic, unsafe and environmentally unfriendly, but above all unable to answer the sophisticated problems posed by global trade in food commodities.

Traditional methods for gross composition analysis (oil, moisture, protein, starch, fibre, ash) are operationally defined, multistep procedures which are carefully described by ISO, BS or AOAC methods for particular types of agricultural materials (AOAC, 1995). Prior knowledge of the description of the material is necessary before the most appropriate method can be selected. Even so, a by-product may be analysed by any one of several equally valid methods with varied results. If the substance presented for testing is unknown or not declared to the analyst it is not possible to select the appropriate method. Wet chemical analyses cannot identify a material and no amount of skill and care by the analyst can compensate for a method which is inadequate for the material tested. For example, lipids covalently linked to polar molecules (e.g. milk fat) will not extract into non-polar solvents without prior acid hydrolysis. The Feedingstuffs Regulations (1985) introduced two methods for the determination of lipid described as method A and B (Cooke, 1986). Method B involving prior hydrolysis was introduced to cope with bound lipids as in feeds containing milk fat.

Instrumental analysis (AAS and ICP) have revolutionised mineral and trace element analysis, while chromatography (GLC, HPLC) allows a vast range of major and trace organics to be measured reliably in feeds. Near infrared (NIR) spectroscopy has become the method of choice for gross composition analyses. However this misses the real point. Traditional methods fail to identify materials and in testing for the obvious, there is a danger of distracting attention from more serious flaws in a raw material which could, and frequently do go undetected.

NIR spectra provide a composition signature, easily acquired in real time, which is vastly more information rich than traditional analyses.Qualitative product identification can be done from a spectra library on computer and the multivariate distance (H statistic or Mahalanobis distance) from the centroid of a population of similar materials can be calculated. This can show if the unknown material tested is typical of its kind; how close it is to the average and how many nearest neighbours it has. A sample having an H value of over 4.0 warrants careful scrutiny. Any larger H statistics should ring alarm bells without *any* reference to traditional measures of composition. As experience of a material is gathered through scanning large numbers of similar samples, the data accumulated becomes a knowledge base which becomes valuable within itself to seek clusters of samples having a common trait such as moulding, soiling, overheating or unspecified contamination or adulteration.Spectra used in this way are ideal for monitoring and QC surveillance where, for the most part, most materials do not warrant expenditure of effort on reference analyses. Hammond (1995) describes this type of quality control as "conformity analysis".

Too much emphasis has been placed on using NIRS merely to replace reference methods having well known inadequacies.Spectra give a global signature of composition which advanced pattern recognition software can use to flag-up hazards not detected at all by traditional tests. Used in this way NIRS can overcome many weaknesses of reference methods.Detection of heat damaged protein (Atanassova and Todorov, 1992; Cho *et al* 1992) and fungal contamination (Roberts, 1992) are examples of novel applications. Product authentication and detection of adulteration are being done by NIRS (Evans *et al*, 1993; Pedretti *et al*, 1993).

In an experiment to evaluate the application of NIRS in the routine analysis of ruminant compound feedstuffs in a commercial mill, three similar sample sets of approximately one hundred samples each were assembled from the routine quality control of the production line (Curtis, 1995). The samples were chosen to have as wide a range as possible in the parameters measured. These parameters were moisture, ash, crude protein, oil A (Soxhlet) and oil B (acid hydrolysis, Soxhlet), and crude fibre measured on an "as received" basis. The analyses were single determinations conducted by the company within the constraints of their normal laboratory operations. The aim of the experiment was to pilot the introduction of NIRS so that the company could judge the likely outcome of purchase of an instrument.

Spectra were obtained in reflectance on an NIRSystems ™ 6500 monochromator (Perstorp Analytical, UK) over the range 400 to 2500 nm and calibration models were developed using partial least squares (PLS) algorithm on ISI Software

(Infrasoft International, Port Matilda, PA, USA) on a 90 MHz pentium computer (Viglen, UK). Models for each set of approximately 100 samples were validated among each other to judge the reliability of the models. A calibration model was also generated on the combined set of some 300 samples (Curtis, 1995). The learning outcome of this project was that some samples were protein concentrates whose spectra and composition were markedly different from the other compounds and were continually flagged as outliers based on the H statistic (Mahalanobis distance) such that these samples did not belong to the same class. Elimination of these outliers generally improved the standard error of cross validation (SECV) of a particular model. The success of NIR calibration depends on the quality of the chemical analyses and the construction of the calibration sample set. Extreme samples may improve the coefficient of determination ($R^2$) but may not improve the standard error of performance (SEP). There is a trade off between the breadth of a calibration model and its ability to discriminate between very similar samples (Wetherill and Murray, 1987). The ISI software (Shenk and Westerhaus, 1993) allows the operator to examine the pattern of sample clusters by plotting principal component scores of spectra in 3D space on the screen to identify global outliers or grouped samples thus assisting calibration set construction using ISI SELECT algorithm. (Figure 6.3).

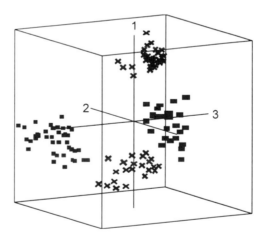

**Figure 6.3** Discrimination of materials using the first three principal components of NIR spectra. Similar materials cluster into groups allowing qualitative identification and discrimination. The distance of any one sample from the centroid of a cluster characterises its closeness to the average of its kind, while distances to its nearest neighbours tell if the sample is unique or if there are others showing the same trait.

Table 6.9 summarises the results obtained in this study. In particular it is noted that the moisture values provided did not calibrate well because the sample moisture had changed between the time of measurement and scanning. This was improved considerably by conducting oven dry matter determinations immediately after scanning. The declaration of analyses on an "as received" basis as in this case leads to very real problems in the expression of results. In this study a significant proportion of the unexplained variance arises from this unsatisfactory situation. It was difficult to obtain calibrations for ash or crude fibre with some samples showing up as persistent outliers warranting repeat analyses (t statistic outliers) or rejection on the grounds of having abnormal spectra (H statistic outliers). Considering however that this sample set consisted of all kinds of ruminant compounds from milk based calf feeds to ewe cobs, the performance could almost certainly be improved by further development work and more careful attention to moisture content and analytical quality control.

**Table 6.9**  NIR CALIBRATION FOR COMPOSITION OF RUMINANT COMPOUND FEEDS

DESCRIPTIVE STATISTICS OF WET CHEMICAL ANALYSIS PROVIDED

|  | *n* | *Mean %* | *Range %* | *SD%* | *CV%* |
|---|---|---|---|---|---|
| Moisture | 309 | 10.86 | 6.74 - 13.6 | 1.10 | 10.1 |
| Ash | 307 | 10.85 | 6.33 - 18.13 | 2.87 | 26.45 |
| Crude Protein | 308 | 19.13 | 12.68 - 48.24 | 6.48 | 33.87 |
| Oil A | 171 | 4.85 | 2.30 - 8.05 | 1.16 | 23.92 |
| Oil B | 241 | 5.15 | 2.50 - 8.60 | 1.34 | 26.02 |
| Crude Fibre | 214 | 9.05 | 1.60 - 15.60 | 2.67 | 29.5 |

NIR CALIBRATION STATISTICS USING PLS

|  | *n* | *Outliers* | *Terms* | *SEC* | $R^2$ | *SECV* |
|---|---|---|---|---|---|---|
| Moisture | 309 | 6 | 6 | 0.63 | 0.66 | 0.66 |
| Ash | 307 | 6 | 6 | 0.62 | 0.95 | 0.69 |
| Crude Protein | 308 | 7 | 10 | 0.76 | 0.99 | 0.82 |
| Oil A | 171 | 7 | 10 | 0.31 | 0.93 | 0.40 |
| Oil B | 241 | 7 | 4 | 0.40 | 0.91 | 0.41 |
| Crude Fibre | 214 | 4 | 7 | 1.01 | 0.86 | 1.14 |

Curtis (1995)

Verheggen *et al* (1990) applied NIR reflectance to 150 ruminant compound feeds used in Belgium. The results (Table 6.10) were adequate except for ash.

These authors claim the SEP of NIRS analyses, for all constituents except ash, are comparable to the between laboratory variation observed with conventional methods in a ring test. The level of accuracy of the measured GE and ME values calculated from the prediction equation of De Boever *et al* (1986) using cellulase digestibility of organic matter and ether extract were satisfactory.

**Table 6.10**   STATISTICS OF NIR CALIBRATION AND PREDICTION FOR RUMINANT COMPOUND FEEDS

|  | *n* | *Range* | *SEC* | *SEP* | $R^2$ |
|---|---|---|---|---|---|
| Crude Protein % | 120 | 11.6 - 40.5 | 0.59 | 0.74 | 0.98 |
| Crude Fibre % | 120 | 6.5 - 19.0 | 0.91 | 0.74 | 0.94 |
| Ether Extract % | 120 | 1.5 - 8.9 | 0.29 | 0.35 | 0.92 |
| Ash % | 84 | 5.7 - 14.9 | 0.94 | 1.03 | 0.45 |
| Cellulase OMD % | 102 | 73.5 - 93.6 | 1.14 | 1.83 | 0.79 |
| Gross Energy (MJ/kg) | 120 | 17.1 - 19.5 | 0.26 | 0.25 | 0.61 |
| Metabolisable Energy (MJ/kg) | 102 | 10.0 - 13.2 | 0.20 | 0.30 | 0.67 |

Verheggen *et al* (1990)

Near Infrared can be applied in a reflectance or transmission mode. Büchmann (1995) demonstrated the use of a transmission analyser (Tecator Infratec 1275) to determine protein and fat in cattle, pig and poultry feeds used in Denmark (Table 6.11).

**Table 6.11**   ACCURACY OF NIT PREDICTIONS IN COMPOUND FEEDS

|  | *n* | *Range* | *SEP* |
|---|---|---|---|
| CATTLE |  |  |  |
| Protein | 40 | 14.6 - 35.4 | 0.62 |
| Fat | 40 | 3.9 - 13.8 | 0.47 |
| Fibre | 38 | 5.8 - 13.7 | 0.68 |
| Moisture | 40 | 9.6 - 14.8 | 0.48 |
| PIG |  |  |  |
| Protein | 109 | 15.4 - 39.1 | 0.62 |
| Fat | 109 | 4.4 - 11.5 | 0.29 |
| Fibre | 80 | 3.8 - 8.6 | 0.71 |
| Moisture | 109 | 8.8 - 13.9 | 0.29 |
| POULTRY |  |  |  |
| Protein | 77 | 15.3 - 25.0 | 0.43 |
| Fat | 77 | 2.8 - 14.4 | 0.32 |

Büchmann (1995)

De Boever *et al.*, (1993) applied NIR to 184 ruminant compound feeds evaluated for composition and *in vivo* digestibility. The set was split equally into a calibration and validation set. Although successful calibrations were obtained for moisture, digestibility and ME value, the errors for crude protein, fat, fibre and ash were considered too high and were caused by incompatibility in the validation set. They obtained better results from a more typical set of real samples from the Belgian National Feed Inspection Service using an enzymatic digestibility to estimate ME. These authors also successfully calibrated an NIR filter instrument with some 800 samples of 18 raw materials for ME calculated from enzymatic digestibility. As a foot note these authors point out that 42 out of 45 independent raw materials could be recognised using a six wavelength discriminant technique. This latter point is important because it shows that the qualitative identification of raw materials is easily achieved by NIRS whereas traditional analysis is unable to identify materials. Three steps in feed evaluation continually get confused. The first step is to ensure that a feed material offered for sale is what it is stated to be (i.e. Qualitative analysis). The second step is to qualify its ranking among like materials (i.e. Quantitative analysis). The third step is to predict the likely animal performance which may be expected from feeding the material to a target animal species (nutritional analysis). All 3 steps are important in different contexts and may apply to both inputs as well as compounded outputs. Spectral database and software can perform all three steps in feed evaluation.

Moya *et al* (1995) applied NIRS to the quality control of 14 raw feed materials for 6 properties. In addition to developing useful calibrations they matched a network of five monochromators to produce similar spectra so that all instruments shared common PLS calibrations. Population structuring using spectra alone (ISI SELECT algorithm) was used to cut the effort of reference analysis. They conclude that NIRS introduced over 8 months, is a significant step forward from their existing QC system. Dardenne *et al* (1995) describe a process for reducing reference analyses required for NIR calibration. Using only 11 samples scanned at varying temperatures, moisture contents and particle sizes, a robust PLS calibration was obtained for protein in wheat. The need to use large numbers of calibration samples to obtain a robust calibration has been a source of disenchantment with NIR. By combining population structuring with this technique it may be possible to reduce analytical effort and make NIR calibration accessible to very labour intensive tests such as *in vivo* nutritional measurements done on just a few carefully selected samples.

Manufacture of consistent quality compound feeds from numerous highly variable ingredients needs frequent if not continuous sampling and instantaneous testing of inputs and finished products. In many industries NIR is well established for process control (e.g. Martens *et al*, 1992). Jensen (1993) describes such a continuous least cost feed formulation system in which up to 8 raw materials are sampled automatically by pneumatic lines. These conduct raw materials to a grinder and NIR device (Figure 6.4).The spectra are then used in control chart logic software to create a dynamic calibration which regularly up-dates the recipe based on raw material and finished product spectra monitored on a repetitive cycle. The process claims to save 0.5 to 2.0% of raw material costs. In another simpler system NIR checks the product mix and "tops-up" the protein level to specification with high grade protein. Closed loop control is an achievable goal in feed manufacturing by making best use of what happens to be in the storage bins. However it does not eliminate the need to check QC at intake to intercept and reject unsatisfactory consignments.

In a laboratory environment sample preparation for NIR becomes a chore which can be done by robotic devices. Eigenraam and Vedder, (1991) and Blank and Vedder (1992) describe an automated system to scan 80,000 forage and feed samples per year in the Netherlands. Whole grain analysers such as the Infratec 1225 (Perstorp Analytical) and Grainspec (Foss) avoid the need to grind cereals for analysis. Small pelleted feeds or inhomogeneous coarse mixes can be tested intact by integrating the spectra of a sufficiently large sample aliquot. Shenk (pers comm) compared the performance of NIR on ground (G) and unground(UG) poultry feeds in a monochromator (NIRSystems 5000) and a 19 filter instrument (Table 6.12). Scanning unground pellets resulted in modest loss in performance against considerably reduced labour.

**Table 6.12**　COMPARISON OF MONOCHROMATOR (MONO) AND FILTER INSTRUMENT PERFORMANCE WITH GROUND (G) AND UNGROUND (UG) MIXED FEED SAMPLES

|  | Range % | SEP MONO (G) | $R^2$ | SEP MONO (UG) | $R^2$ | SEP FILTER (G) | $R^2$ |
|---|---|---|---|---|---|---|---|
| Crude Protein | 14.0 - 34.0 | 0.55 | 0.98 | 0.80 | 0.95 | 0.86 | 0.96 |
| Fat | 2.0 - 13.0 | 0.25 | 0.99 | 0.56 | 0.94 | 0.40 | 0.97 |
| Dry Matter | 85.0 - 93.0 | 0.30 | 0.98 | 0.35 | 0.98 | 0.35 | 0.96 |

Shenk, pers comm (1995)

## CONCLUSION

The feed industry faces increasing legislation and regulation in the conduct of its operations. Quality control places increasing burdens on laboratories serving the industry. Producing consistent quality compound feeds from highly variable raw materials requires continuous sampling. Traditional methods of analysis are

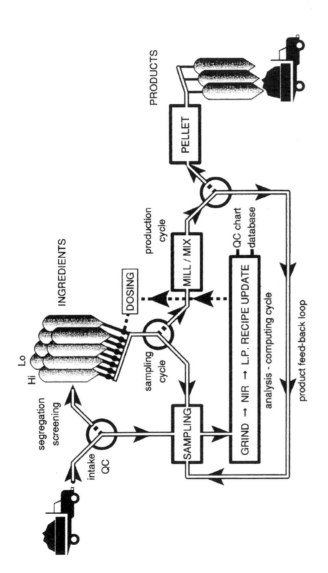

**Figure 6.4** Closed loop feed-back control in feed manufacture using NIR

Aliquots of each ingredient are automatically sampled from silos by pneumatic lines sent to an NIR analysis unit which computes the least cost recipe and doses ingredients to the mill for mixing and pelleting. Product is likewise sampled and analysed and the recipe continuously updated by frequent sampling cycles during manufacture. (Modified after the PRINACON system described by Jensen, 1993)

fragmentary and do not answer industrial need because labour costs and delay in reporting make such methods retrospective. Furthermore official methods are for the most part multistep, operationally defined procedures requiring prior knowledge of the definition of the material tested to allow appropriate method selection. Trade, tarriffication and traceability are compromised by this lack of definition of materials, leaving loopholes in legislation.

Near infrared spectroscopy has been in use for nearly 20 years during which time applications have grown in sophistication from simple filter instruments performing a restricted range of tests on well defined materials such as moisture, protein and oil in cereals and oilseeds, to complex analyses of composition and animal performance attributes in wet forage using monochromators and specialist software (Barber *et al*, 1990). While NIRS has had an enormous economic impact, its full scope is only beginning to be realised as a method for classifying and characterising highly variable organic materials and optimising their use as feed resources. Reliability of instruments and vastly better software and computers have opened up new applications in a wide range of industries. Multivariate statistical methods such as Principal Component Analysis (PCA) and Partial Least Squares (PLS) algorithms allow materials to be classified from spectra alone prior to analysis using large spectral databases which are currently being collected for CD ROM. NIR calibrations or pre calibrated instruments can be purchased and immediately pressed into service. NIRS will remain a secondary method of analysis dependent on classical methods as the primary standard for quantitative measurements. However harmonisation of testing is more likely to be achieved through the use of NIR monochromators which can be spectrally matched and networked to share common calibration bases (Dardenne *et al*, 1992). Leadership from regulatory agencies in effecting this process multinationally is long overdue.

The successful application of NIR to analysis of numerous raw materials and compounds has been adequately demonstrated in publications and NIRS has been in regular use by a few major multinationals for some years although performance data are seldom in the public domain. Least cost ration formulation can be conducted on the measurements derived from spectra or from spectral data itself. Many industries use NIR on-line, some with closed loop feedback control to manage the manufacture of consistent product continuously. While these systems are more common and more easily installed in petrochemical, plastics, and pharmaceutical industries, they are being applied to feed compounding by continuous least cost formulation or "topping-up" systems to achieve uniform products from highly variable ingredients.

While NIRS may not determine trace constituents in feeds, nor may it provide a universal solution to feed testing methodology, it will become the method of choice for routine monitoring of quality and consistency of inputs and products for a wide range of properties in a timely and cost effective manner. We are led to believe that quality surveillance will either be done by NIRS or it will not be done at all. The next decade may prove that the "do-nothing" option is not an option at all for those who wish to stay in business.

# References

Alderman, G. (1985). Prediction of the energy value of compound feeds. In: *Recent Advances in Animal Nutrition - 1985,* pp. 3-52. Edited by W. Haresign and D. J. A. Cole, London: Butterworths.

AOAC (1995). Official Methods of Analysis, 16th Edition, AOAC, Arlington, VA, USA.

Atanassova, S. and Todorov, N. (1992). The use of NIRS for prediction of heat damaged protein in Alfalfa (Medicago sativa) forages. In: *Making Light Work: Advances in NIRS.* Edited by I. Murray and I. A. Cowe, VCH Weinheim pp 314-317.

Ball, J. A. (1988). Variability of raw materials. In: *The Feed Compounder.* **Jan.** pp. 18-23. H. G. M. Publications, Derbyshire.

Barber, G. D., Givens, D. I., Kridis, M. S., Offer, N. W. and Murray, I. (1990). Prediction of the organic matter digestibility of grass silage. *Animal Feed Science and Technology,* **28**, 115-128.

Blank, F. T. and Vedder, H. W. (1992). Automated NIR analysis for routine feed evaluation. In: *Near Infrared Spectroscopy - Bridging the Gap between Data Analysis and NIR applications.* Edited by K. I. Hildrum, T. Isaksson, T. Naes and A. Tandberg, Ellis Horwood, Chichester, UK pp 197-202.

DeBoever, J. L., Cottyn, B. G., Vanacker, J. M. and Boucqué, Ch. V. (1993). Recent developments in the use of NIRS for evaluating compound feeds and raw materials for ruminants. Paper 1 *Proceedings Conference NIR Spectroscopy - Developments in Agriculture and Food*, Birmingham, UK. ADAS Drayton Res Centre FEU, 8pp.

DeBoever, J. L., Cottyn, B.G., Buysse, F. X., Wainman, F. W. and Vanacker, J. M. (1986). The use of an enzymatic technique to predict digestibility, metabolisable and net energy of compound feedstuffs for ruminants. In: *Anim. Feed Sci. Technol.* **14**, pp. 203-214.

Brigstocke, T. (1989). Variability of Raw Materials. *The Feed Compounder*, Sept, pp 65-66. Edited by H. G. Mounsey, H. G. M. Publications, Derbyshire.

Büchmann, N. B. (1995). The use of a new near infrared transmission (NIT) analyser from Tecator for determining protein and fat in animal feeds. In: *Leaping Ahead with Near Infrared Spectroscopy.* Edited by G. Batten, P.

C. Flinn, L. A. Welsh and A. B. Blakeney, NIRS Group, Royal Australian Chemical Institute, Melbourne, pp 248-251.

CEC. (1994).  The circulation of feed materials.  COM(94) 313 final, Brussels 20.7.94

Cho, R. K. Ozaki, Y., Ahn, J. J. and Iwamoto, M. (1992).  The application of near infrared reflectance spectroscopy for monitoring changes in secondary structure of denatured proteins.  In: *Near Infrared Spectroscopy - Bridging the Gap between Data Analysis and NIR Applications.*  Ellis Horwood, Chichester, UK, pp 333-338.

Chung, D. S. and Pfost, H. B. (1964).  Overcoming the effects of ingredient variation.  In: *Feed Age* **14(9)**, pp. 24-27.

Cooke, B. C. (1986).  The implications to research and the feed compounder of the new oils and fats determination.  In: *Recent Advances in Animal Nutrition - 1986,* pp. 83-86.  Edited by W. Haresign and D. J. A. Cole, London: Butterworths.

Curtis, M. (1995).  An investigation into the application of NIRS to the Quality Control of Ruminant Compound Feeds, *MSc Thesis*, Aberdeen University, 57pp.

Dardenne, P., Biston, R. and Sinnaeve, G. (1992).  Calibration transferability across NIR instruments.  In: *Near Infrared Spectroscopy - Bridging the Gap between Data Analysis and NIR Applications.* Edited by K. I. Hildrum, T. Isaksson, T. Naes and A. Tandberg, Ellis Horwood, Chichester, UK, pp 453-458.

Dardenne, P., Sinnaeve, G. Bollen, L. and Biston, R. (1995).  Reduction of wet chemistry for NIR calibrations.  In: *Leaping Ahead with Near Infrared Spectroscopy*, Edited by G. Batten, P. C. Flinn, L. A. Welsh and A. B. Blakeney, NIRS Group, Royal Australian Chemical Institute, Melbourne, pp 154-160.

Duncan, M. S. (1988).  Problems of dealing with raw ingredient variability.  In: *Recent Advances in Animal Nutrition - 1988,* pp. 3-11.  Edited by W. Haresign and D. J. A. Cole, London: Butterworths.

Eigenraam, A. and Vedder, H. W. (1991).  Autosampler for NIR routine analysis.  In: *Proceedings 3rd International Conference NIR*, Brussels, Edited by R. Biston and N. Bartiaux-Thill, Ag. Res. Centre, Gembloux, Belgium, pp 30-33.

Evans, D. G., Scotter, C. N. G., Day, L. Z. and Hall, M. N. (1993).  Determination of the authenticity of orange juice by discriminant analysis of Near Infrared spectra: A study of pretreatment and transformation of spectral data. *J. Near Infrared Spectroscopy* **1**, 33-44.

Hall, A. (1983).  The use of Near Infrared (NIRS) Reflectance Spectrocomputer for analyses of compound feeds, *MSc thesis*, University of Aberdeen, 142pp.

Hammond, S. V. (1995). The cost avoidance role of NIR in pharmaceutical production. In: *Leaping Ahead with Near Infrared Spectroscopy*, Edited by G. Batten, P. C. Flinn, L. A. Welsh and A. B. Blakeney, NIRS group, Royal Australian Chemical Institute, Melbourne, pp 394-404.

Henneberg, W. and Stohmann, F. (1864). Beiträge zur begründung einer rationellen. Fütterung der WeiderKäuer II Braunschweig, **29**, 48.

Jensen, E. P. (1993) Use of NIR for on-line process control during the manufacture of animal feedstuffs. Paper 2 Proceedings Conference *NIR Spectroscopy - Developments in Agriculture and Food*, Birmingham, UK, ADAS Drayton Res. Centre, FEU, pp 8.

Kjeldahl, J. (1883). Zeitschrift für analytische chemie, **22**, 366.

MAFF (1993). Prediction of energy values of compound feedingstuffs for farm animals - *A guide for farmers*. Leaflet PB1272, MAFF Publications, London.

Martens, A. Cermelli, I. Descales, B., Llinas, J. R., Vidal, J. L. and Margail, G. (1992). NIR process control of a steam cracker. In: *Making Light Work: Advances in NIR Spectroscopy*, Edited by I. Murray, I. A. Cowe, VCH, Weinheim, pp 477-481.

Moya, L., Garrido, A., Guerrero, J. E., Lizaso, J. and Gómez, A. (1995). Quality control of raw materials in the feed compound industry. In: *Leaping Ahead with Near Infrared Spectroscopy*, Edited by G. Batten, P. C. Flinn, L. A. Welsh and A. B. Blakeney, NIRS Group,, Royal Australian Chemical Institute, Melbourne, pp 111-116.

Murray, I. and Hall, A. P. (1983). Animal feed evaluation by use of Near Infrared Reflectance (NIR) Spectrocomputer. *Analytical Proceedings*, **20**; 75 - 79.

Murray, I. (1986). Near Infrared Reflectance analysis of forages. In: *Recent Advances in Animal Nutrition - 1986,* pp. 141-156. Edited by W. Haresign and D. J. A. Cole, London: Butterworths.

Pedretti, N., Bertrand, D., Semenou, M., Robert, P. and Giangiacomo, R. (1993). Application of an experimental design to the detection of foreign substances in milk. *J. Near Infrared Spectroscopy* **1**, 174-184

Roberts, C. A.. (1992). Quantification of fungal contaminants in forage and grain. In: *Making Light Work: Advances in NIR Spectroscopy*, Edited by I. Murray, and I. A. Cowe, VCH Weinheim, pp 352-356.

Shenk, J. S. and Westerhaus, M. O. (1993). Analysis of Agriculture and Food Products by Near Infrared Reflectance Spectroscopy. *Monograph*, Infrasoft International, Port Matilda, PA USA.

Statutory Instruments (1982,84,85). Feedingstuffs (Sampling and Analysis) Regulations, 1982/1144; 1984/52; 1985/1119.

UKASTA, ADAS and COSAC (1985). Prediction of energy values of compound feed. *Report Working Party*, MAFF, London.

108 *Traditional analytical methods and near infrared spectroscopy*

Verheggen, J., Dardenne, P. Théwis, A. and Biston, R. (1990). Application of near infrared reflectance spectroscopy (NIRS) for predicting chemical composition, cellulase digestibility of organic matter and energy values of compound feedstuffs for ruminants. In: *Proceedings 3rd International Conference on NIRS*, Brussels, Edited by R. Biston and N. Bartiaux-Thill, Agricultural Centre, Gembloux, Belgium, pp 287-290.

Wainman, F. W., Dewey, P. J. S. and Boyne, A. W. (1981). Feedingstuffs Evaluation Unit, Rowett Research Institute, *3rd Report*, Dept Ag. Fisheries Scotland, Edinburgh.

Wetherill, G. Z. and Murray, I. (1987). The spread of the calibration set in near-infrared reflectance spectroscopy. *J. Agr. Sci. (Cambridge)*, **109**: 539-544.

Williams, D. R. (1987). Animal Feeding Stuffs Legislation of the U.K. - A concise guide, London: Butterworths. 135pp.

Windham, W. R. and Flinn, P. C. (1992). Comparison of MLR and PLS regression in NIR analysis of quality components in diverse feedstuff populations. In: *NIRS - Bridging the Gap between Data Analysis and NIR Applications*. Edited by K. I. Hildrum, T. Isaksson, T. Naes and A Tandberg, Ellis Horwood, Chichester, UK, pp 459-464.

# Appendix

Notation:

| | | |
|---|---|---|
| Y | = | actual reference method values |
| $\bar{Y}$ | = | mean of actual reference method values |
| $\hat{Y}$ | = | NIR predicted values |
| $n$ | = | number of samples in the model |
| $p$ | = | number of terms in the model |

Mean: $\bar{Y} = \dfrac{\Sigma Y}{n}$    Standard Deviation: $SD = \sqrt{\dfrac{\Sigma(Y - \bar{Y})^2}{n-1}}$

Total Sum of Squares $\qquad\qquad SS_{TOT} = \Sigma(Y - \bar{Y})^2$

Sum of Square for Residuals $\quad SS_{RES} = \Sigma(Y - \hat{Y})^2$

Sum of Squares for Regression $SS_{REG} = SS_{TOT} - SS_{RES} = \Sigma(\hat{Y} - \bar{Y})^2$

MEAN Square for Regression $\quad MS_{REG} = \dfrac{SS_{TOT} - SS_{RES}}{\text{d.o.f. regression}} = \dfrac{\Sigma(\hat{Y} - \bar{Y})^2}{p+1}$

MEAN Square for Residuals $\quad MS_{RES} = \dfrac{SS_{RES}}{\text{d.o.f. residuals}} = \dfrac{\Sigma(Y - \hat{Y})^2}{n-p-1}$

F TEST STATISTIC $= \dfrac{MS_{REG}}{MS_{RES}} = t^2$

$t$ TEST STATISTIC $= \dfrac{\text{Residual}}{SEC} = \dfrac{(Y - \hat{Y})}{SEC} = \sqrt{F}$

Samples having $t > 2.5$ are outliers which warrant repeat reference analysis

Coefficient of Multiple Determination: fraction of variance explained

$$R^2 = 1 - \left(\frac{SEC}{SD}\right)^2 \cong \frac{SS_{REG}}{SS_{TOT}} = \frac{SS_{TOT} - SS_{RES}}{SS_{TOT}}$$

Standard Errors of Estimate: calibration, prediction, cross validation

$$SEE, SEC, SEP, SECV = \sqrt{MS_{RES}}$$

Calibration Set:                   Validation Set:

$$SEC = \sqrt{\frac{\Sigma(Y - \hat{Y})^2}{n-p-1}} \qquad SEP\ or\ SECV = \sqrt{\frac{\Sigma(Y - \hat{Y})^2}{n-1}}$$

Prediction Sum of Squares      $$BIAS = \frac{\Sigma Y}{n} - \frac{\Sigma \hat{Y}}{n} = B$$

$$PRESS = \Sigma(Y - \hat{Y})^2$$

Standard Error of Performance
Corrected for BIAS:

$$SEP(C) = \sqrt{\frac{\Sigma(Y - \hat{Y} + B)^2}{n-1}}$$

Standard Error of the Laboratory:

$$SEL = \sqrt{\frac{\Sigma(y - \bar{y})^2}{n}} = \sqrt{\frac{\Sigma(y_1 - y_2)^2}{2n}}\ \text{for duplicates } y_1, y_2, \bar{y} = \frac{y_1 + y_2}{2}$$

**IV**

**Ruminant Nutrition**

7

# MILK ANALYSIS AS AN INDICATOR OF THE NUTRITIONAL AND DISEASE STATUS OF DAIRY COWS

B. PEHRSON
*Experimental Station, Swedish University of Agricultural Sciences, P.O. Box 234, S—532 23 Skara, Sweden*

## Introduction

Blood profiles are commonly used in human medicine for the diagnosis of diseases and to monitor the patients' health. They are valuable because modern techniques make it possible to analyse many blood constituents simultaneously, and because in most cases there are direct and rapid routes of communication between the patients and the analytical laboratory.

Blood profiles can also be used in farm animals. The "Compton Metabolic Profile Test" was introduced by Payne, Dew, Manston and Faulks (1970) and includes a wide range of analyses. Later a "Mini Metabolic Profile Test" was introduced by Blowey (1975), in which the analyses are restricted to glucose, urea and albumin. However, it has been pointed out by, among others, Adams, Stout, Kradel, Guss, Mosel and Jung (1978) and Dirksen (1994) that blood has several practical disadvantages as a test medium.

In contrast, milk should be an ideal medium for evaluating the health and the efficiency of production of dairy cattle, because samples of milk are sent routinely to the dairy's laboratory at least once a month; there would therefore be no extra costs for sampling and transport, and farmers would be encouraged to make the best use of the information which is available so relatively cheaply. In addition, they would be able to make use of the tests carried out on the farm, either on the milk of individual cows or on the bulk milk. It is therefore not surprising that there has been an increasing interest in "Milk Profile Tests" during the last decade. The aim of this paper is to evaluate the potential value of different constituents of milk as biological indicators of the health and productivity of dairy cows.

113

## Somatic cell counts

Mastitis is economically the most important disease of dairy cows. The costs of treating the disease are high, but they are almost negligible in comparison with the losses due to the decreased milk yields of chronically infected cows (Taponen and Myllys, 1995). As a result great efforts are made to try to reduce the incidence of the disease, and to achieve this aim it is vital to identify any cows with chronic mastitis, so as to prevent them from spreading the infective organisms to the healthy cows in the herd.

It is well documented that chronic mastitis is characterised by an increase in the number of somatic cells in the milk. Most of them are white blood cells which have been attracted from the blood into the udder as a defence against the infection. Regular measurements of milk cell counts are therefore valuable for revealing mastitis and every effort should be made to reduce the cell count – both in samples from individual cows and in the bulk milk. Practical measures to reduce the cell count can be encouraged either by reducing the price paid for milk with a high cell count or by paying more to farmers who consistently produce milk with a low cell count. Both these methods are practised in Finland (Saloniemi, 1995), and a bonus is paid when the cell count in the bulk milk is less than 250,000/ml. In Sweden there is a price reduction for bulk milk containing more than 400,000 cells/ml, and if the cell count exceeds this value repeatedly during a period of six months the farmer is not allowed to deliver any more of the milk to the dairy. As a result farmers tend to cull cows with chronic mastitis rather than keeping them; furthermore, an important positive psychological effect is induced by the farmers' awareness of the importance of good udder health in their herds.

## Fat concentration

Fat is the solid constituent of milk whose concentration is most easily affected by nutritional means, and the "low milk fat syndrome" has been studied for many years. The syndrome is not a disease but a metabolic consequence of trying to achieve a high milk yield as economically as possible. To make full use of the modern cow's genetic capacity for high milk production it is necessary to feed a ration which has a much higher energy concentration than the natural diet of a ruminant animal. Such a ration carries the risk of reducing the milk fat concentration, particularly if the ratio of forage/concentrate in the ration is low, or its concentration of acid-digestible fibre (ADF) is low, or the total dietary intake of starch is high (Sutton, 1989). The risk increases when the forage is of high nutritional quality and finely chopped, when the diet contains unsaturated fatty acids, when the concentrates are rolled or ground, or when the concentrates are fed only twice a day. The changes induced in the metabolism of the rumen by such diets tend to reduce the rumen pH, increase the production of propionic acid, reduce the ratio of acetate/propionate, and reduce the numbers of protozoa (Storry, 1970; Engvall, 1980).

However, a moderate reduction in the percentage of fat in the milk can be completely compensated by an increase in milk yield (Figure 7.1), and in these circumstances there is normally no need to change the diet. Whether a reduction in fat concentration may have a negative economic effect will depend to a great extent on the principles which determine the payments for milk in a particular country; it will also depend on whether the reduced fat concentration is accompanied by a reduction in protein concentration – as has been observed in field cases by Engvall (1980) – or by an increase in protein concentration – as has been observed experimentally by Emery (1978).

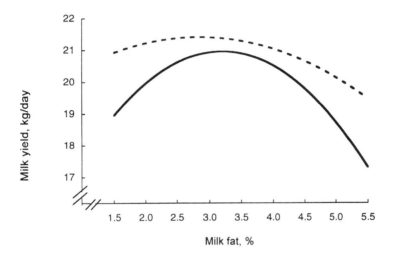

**Figure 7.1** The relation between daily milk production and milk fat concentration in Swedish Red and White cows (---) and Swedish Friesian cows (—) during the 4th month of lactation (from Engvall, 1980)

It is important to be aware that there may be a considerable time lag between when a diet which is liable to reduce the milk fat concentration is introduced and when the reduction occurs; Engvall (1980) found that the interval until the milk fat decreased below 2.0% varied between 34 and 65 days. However, the decrease occurred suddenly (Figure 7.2), and was accompanied by significant changes in the ruminal microbiota.

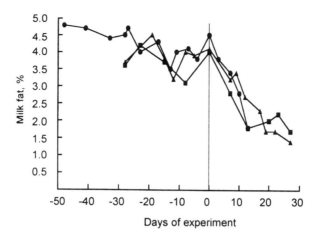

**Figure 7.2** The time lag from the start of a milk fat depressing diet (2 kg long hay, 2 kg hay wafers and 10–13 kg pelletted concentrates) to the onset low milk fat syndrome in three cows (from Engvall, 1980)

## Protein concentration

Changes in diet are known to have a much smaller effect on milk protein concentration than on milk fat, but nevertheless milk protein concentrations can give some indication of the efficiency of a diet. Emery (1978) concluded that increases in milk protein concentration could be achieved by feeding more energy, more crude protein and less fibre, and more recent reviews (Sutton, 1989; Spörndly, 1989) have concluded that there is a positive correlation between energy intake and milk protein concentration. Spörndly analysed the results of 53 feeding experiments and calculated that on average an increase of 1 MJ in the daily intake of metabolisable energy could be expected to increase the milk protein content by between 0.003 and 0.005 percentage points. She also concluded that the effects of the quantity of concentrates fed and of a low fibre diet were indirect effects, and that dietary fat (ether extract) had a negative effect on the protein content of milk. Furthermore, roughage in the form of hay seemed to result in higher milk protein concentrations than silage.

Severe underfeeding with protein can reduce the protein concentration of milk (Gordon, 1977), but the effects of feeding a surplus of protein appear to be variable; Emery (1978) reported a positive effect from feeding more crude protein, but Sutton (1989) considered that extra dietary protein, whether rumen-degradable or not, had no significant effect on milk protein concentration, and the results presented by Spörndly (1989) are mainly in agreement with that conclusion.

## Ratio of fat/protein

Hagert ( 1991) and Dirksen (1994) proposed that the ratio of the concentrations of fat and protein in milk (FP ratio) could be used as an indicator of the dietary energy balance of a cow; the proposal was based on the following sequence of metabolic events:

An energy deficit during peak lactation results in the mobilisation of lipids from the body fat depots, an increase in the blood concentration of free fatty acids, and an increase in the production of fat by the udder. At the same time the energy deficit in the rumen reduces the rate of synthesis of bacterial protein; the supply of amino acids to the udder is reduced and there is a reduction in the protein concentration in the milk. They studied cows which were fed energy according to their requirement, or above or below their requirement, and found that their FP ratios provided a moderately good indication of their dietary energy balance. A FP ratio less than 1.4 indicated an optimal or positive energy balance, and a ratio above 1.4 indicated en energy deficit. During peak lactation many energy deficient cows had FP ratios above 2.1.

## Lactose concentration

Severe underfeeding can result in a reduction in the lactose content of milk. There are also a few reports that a low ratio of forage/concentrate in the diet can increase milk lactose, and that high fat supplementation can cause a decrease (Sutton, 1989). However, although these changes are statistically significant, they are so small that milk lactose is of no practical value as an indicator of diseases or metabolic disturbances.

## Progesterone

An optimal calving interval is vital for economic dairy production. The optimal interval may vary slightly from country to country and from farm to farm, but it should not be very far from 365 days; it is widely accepted that for every day by which the calving interval exceeds 365 days there will be a cost of approximately two pounds. It is therefore vitally important that cows are inseminated within three months of calving and at the correct time during an ovarian cycle.

By convention, each ovarian cycle (Figure 7.3) begins with an ovulation, which is defined as the release of an egg from a follicle located on the surface of one of the two ovaries. Within a few days the follicle is then transformed into a corpus luteum which secretes the hormone progesterone; during this luteal phase of the cycle high levels of progesterone are secreted into the milk where it can be detected by a suitable laboratory or on-farm test. In a normally cycling, non-pregnant cow the progesterone concentration in milk reaches a peak value about 10 days after ovulation, remains at this level for four to five days, and then decreases rapidly until 17 to 18 days after ovulation when it is barely detectable in the blood or the milk. The corpus luteum regresses, and approximately 21 days after the preceding ovulation a new follicle releases an egg to complete the ovarian cycle. The cow will be in oestrus one to two days before each ovulation.

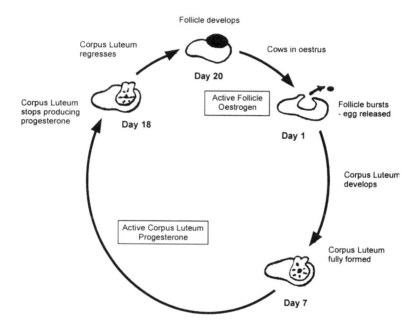

**Figure 7.3** The ovarian cycle of a dairy cow

If an egg is fertilized the corpus luteum does not regress and it continues to produce progesterone throughout the pregnancy; as a result the concentration of progesterone in the cow's milk will still be high about 19 days after it was inseminated. A progesterone concentration less than 5 ng/ml indicates the oestrus phase, a concentration more than 10 ng/ml the luteal phase, and intermediate concentrations indicate either luteolysis or the formation of a new corpus luteum (Ahlin and Larsson, 1985).

A measurement of milk progesterone concentration can therefore be used to test whether a cow is pregnant to an insemination. If the progesterone concentration is high in a milk sample taken 19 days after the cow was inseminated the cow is probably pregnant, and the pregancy would be confirmed if the concentration is still high in a sample taken five days later. However, if the concentration is low in either of the samples the cow is certainly not pregnant.

A milk progesterone measurement can also be useful at other times. For example, if the progesterone concentration is high when the cow is inseminated, the cow must have been in the luteal phase and therefore cannot become pregnant. It is also possible to predict the time of an oestrus suitable for the first insemination after calving; three samples should be taken, about one week apart, the first about 40 days after calving. In a normally cycling cow a mixture of high and low progesterone concentrations should be found, and an oestrus can be expected approximately 21 days after the day on which a low concentration was recorded.

## Ketone bodies

In a high-yielding dairy cow the maximal demand for nutrients is reached within two weeks after calving, whereas the cow's voluntary food intake does not reach its maximum until several weeks later (Wiktorsson, 1971; Foster, 1988). As a result the cow is almost inevitably in negative energy balance during at least the first month of lactation. It seems to be a basic biological rule that the body reserves of an animal that has recently given birth should be regulated hormonally so that they can be mobilised to satisfy the nutritional requirements of its offspring. It is certainly no accident that in dairy cows the period of risk for ketosis coincides with the period when the cow's voluntary intake fails to balance its nutritional losses through the milk. As a result at least 98% of all cases of ketosis occur during the first eight weeks of lactation, with a peak incidence between three to five weeks after calving (Øverby, Aas Hansen, Jonsgård and Søgnen, 1974; Simensen, Halse, Gillund and Lutnaes, 1990).

The relationship between negative energy balance and ketosis was established scientifically several decades ago (e.g. Pehrson, 1966; Henkel and Borstel, 1969). However, in order to reduce the incidence of hyperketonaemia it is necessary to take into account not only the energy balance, but also the feeding regimen, as has been established by many authors, among them Gustafsson (1993), who defined some of the important dietary variables.

Besides being due to a deficiency of energy in general, ketosis in high-yielding dairy cows is related to a specific lack of glucose. The carbohydrates in the cow's diet are broken down in the rumen to volatile fatty acids, predominantly acetic and butyric acids, nether of which is glucogenic, and as a result most of the dietary carbohydrates are, from the point of view of glucogenesis, "destroyed" in the rumen. Propionic acid, which is transformed into glucose in the liver, constitutes only 15 to 20% of the volatile fatty acids produced in the rumen.

The quantity of glucose absorbed directly from the gastrointestinal tract of ruminants is negligible, and high-yielding dairy cows are therefore highly dependent on an effective system for producing glucose from other sources ("gluconeogenesis"). The main gluconeogenetic resources available are the propionic acid produced in the rumen, the glucogenic amino acids derived from the ration or mobilised from body protein, the lactate produced in the rumen and by cell metabolism, and the glycerol produced in the rumen or mobilised from body fat.

The importance of an effective level of gluconeogenesis is underlined by the fact that the level of milk production is heavily dependent on the concentration of lactose in the alveolar cells of the udder, because lactose is the major contributor to the osmotic pressure of milk; however, for the synthesis of lactose the udder cells require glucose. A cow with a high genetic capacity for milk production presumably gives its udder preference for the supply of glucose, but if there is a shortage of glucose – which is very likely during the first few weeks of lactation in an energy deficient high-yielding cow – hypoglycaemia will develop and there will be a shortage of oxaloacetate. As a result the fatty acids mobilised from the body fat cannot be used in the citric acid cycle but are transformed into ketone bodies which are secreted from the blood into the milk (Figure 7.4).

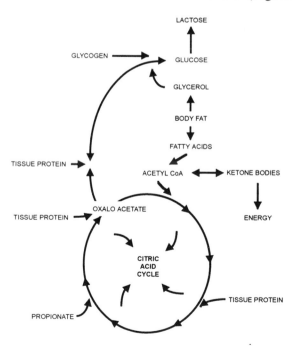

**Figure 7.4** Metabolic pathways for glucose mobilization and lactose production

Plasma glucose concentration is thus one of the factors controlling ketogenesis and is negatively correlated with the concentration of ketone bodies (Herdt, 1988). It has been shown that cows with hypoglycaemia have a much lower pregnancy rate to the first insemination than cows with normal or high plasma glucose levels (Plym Forshell, Andersson and Pehrson, 1991; Pehrson, Plym Forshell and Carlsson, 1992; Table 7.1). At that time the risk period for ketosis has passed, but Plym Forshell *et al.* (1991) found that the cows with low fertility had been hypoglycaemic three to seven weeks after calving and were also hyperketonaemic three to five weeks after calving (Table 7.2). It should therefore be possible to evaluate the risk of a reduction in fertility by measureing either plasma glucose or milk acetone a few weeks after calving; because of the practical problems in using blood, the measurement of milk acetone is much to be preferred. There is high correlation between the concentrations of ketone bodies in blood and acetone in milk; the abscence of significant diurnal variation even makes measurements milk acetone more suitable than measurements of blood ketone bodies (Andersson and Lundström, 1985).

**Table 7.1** THE NUMBERS AND PERCENTAGES OF COWS WHICH BECAME OR DID NOT BECOME PREGNANT AT THE FIRST INSEMINATION, IN RELATION TO THEIR PLASMA GLUCOSE CONCENTRATIONS (PEHRSON *et al.*, 1992).

| *Plasma glucose* | *Pregnant* | | *Not pregnant* | |
|---|---|---|---|---|
| *mmol/l* | *n* | *%* | *n* | *%* |
| < 2.50 | 22 | 36.1 | 39 | 63.9 |
| 2.50–2.99 | 40 | 41.7 | 56 | 58.3 |
| 3.00–3.39 | 47 | 48.5 | 50 | 51.5 |
| ≥ 3.40 | 26 | 65.0 | 14 | 35.0 |
| Significance | | p < 0.05 | | |

**Table 7.2** MEAN PLASMA GLUCOSE AND MILK ACETONE CONCENTRATIONS (MMOL/L) EARLIER IN THE LACTATION PERIOD IN COWS CLASSIFIED BY THEIR PLASMA GLUCOSE CONCENTRATION AT THE TIME OF INSEMINATION. n = NUMBER OF COWS (FROM PLYM FORSHELL *et al.*, 1991)

| | *Glucose concentration at insemination* | | | | | |
|---|---|---|---|---|---|---|
| *<2.50* | | | *2.50–3.40* | | *>3.40* | |
| | n | $\bar{x}$ | n | $\bar{x}$ | n | $\bar{x}$ |
| Glucose week 3–4 | 30 | 2.25 | 218 | 2.73 | 35 | 2.90 |
| Glucose week 6–7 | 30 | 2.49 | 217 | 2.74 | 31 | 2.87 |
| Acetone week 3–4 | 31 | 0.52 | 214 | 0.21 | 34 | 0.14 |
| Acetone week 5 | 18 | 0.91 | 179 | 0.20 | 28 | 0.17 |
| Acetone week 6–7 | 34 | 0.21 | 227 | 0.17 | 31 | 0.16 |

Plym Forshell *et al.* (1991) classified the milk acetone concentrations according to the limits suggested by Andersson (1984); concentrations less than 0.41 mmol/l were considered normal, 0.41-1.00 as slightly above normal, 1.01-2.00 mmol/l as moderately above normal and more than 2.00
mmol/l as very high. Most of the cows with levels slightly above normal were not expected to show clinical signs of ketosis, whereas nearly all the cows with very high levels would have been expected to show signs which were apparent to a farmer. Gustafsson (1993) used the same FIA (flow injection analysis) technique for measuring milk acetone, but proposed that concentrations up to 0.7 mmol/l should be regarded as normal. On-farm tests based on sodium nitroprusside can give a fairly good indication of the true acetone concentration to those who have learned to interpret the colour changes with this reagent. Recently, a new test strip has been introduced by Dirksen (1994) for assessing semi-quantitatively the concentration of betahydroxybutyrate in milk. Concentrations less than 0.1 mmol/l are considered normal, and there were quite good correlations with the results of the FIA method used by Andersson (1984) for measuring milk acetone.

The relationship between hyperketonaemia – indicated by the concentration of milk acetone – and reduced fertility has been observed by several other workers. Berglund and Larsson (1983) found that there was a delay in the normalisation of the reproductive functions of hyperketonaemic cows. Andersson and Emanuelsson (1995) recorded prolonged intervals between calving and the first and last inseminations in herds with a high prevalence of hyperketonaemia, and like Refsdal (1982), they also observed an increased risk of cystic ovaries in hyperketonaemic cows. Gustafsson (1993) observed negative effects of hyperketonaemia on fertility only in cows with milk acetone concentrations above 2.0 mmol/l.

Another reason for measuring ketone bodies in milk is that, in addition to their effects on fertility, both subclinical and clinical ketosis can reduce the yield of milk. Lucey, Rowlands and Russel (1986) recorded a reduction of about 5 kg/day a few weeks before clinical signs were observed. Dohoo and Martin (1984), Andersson and Lundström (1985), and Simensen *et al* (1990) also reported reduced milk yields in subclinical cases of ketosis; the reductions were between 4–5% or 3–4 kg /day. Much larger reductions in milk yield have been recorded in cases of clinical ketosis (Andersson, 1988; Gustafsson, Andersson and Emanuelson, 1993; Figure 7.5).

As mentioned by Andersson (1988), measurements of ketone bodies in milk can be used either for detecting hyperketonaemia in individual animals, or for revealing suboptimal dietary regimens in herds with a high incidence of clinical and subclinical ketosis – herds which may need nutritional advice. It may also be possible to use the routine measurements of ketone bodies in the milk of dairy cows to assess the genetic contribution of the sires used for artificial insemination to the susceptibility of their daughters to ketosis.

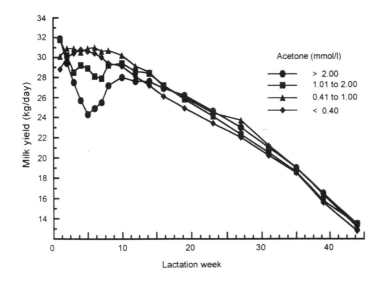

**Figure 7.5** Lactation curves for daily milk yield in multiparous cows in four classes of milk acetone concentration (courtesy by Gustafsson *et al.*, 1993)

## Urea

Diets for dairy cows are calculated on the basis of tables which list the cow's requirements for different nutrients at each stage of lactation, coupled with tables which give the average nutritional composition of different feedstuffs. The calculations are based to a large extent on theoretical models of complicated biological processes, and as a result they are subject to many errors, even when the nutritional composition of each component of a diet has been analysed. It would therefore be extremely valuable if biological indicators, which would reflect the final outcome of these complicated metabolic processes, could be used to complement these theoretically calculated rations in the evaluation of the true adequacy of a diet. The appropriate indicators should preferably be simple and cheap to analyse.

Although it is not ideal, milk urea satisfies several of the criteria for a useful biological indicator. Current knowledge indicates that the concentration of urea in milk is closely related to the balance between the levels of energy and protein in the diet. Furthermore, urea is simple and cheap to analyse in the samples of bulk or individual cow's milk which are sent regularly to the dairy.

The diagram of the metabolic processes which occur in the rumen (Figure 7.6) shows that the amount of ammonia produced in the rumen is dependent on the amount of protein in the diet. This ruminal ammonia can be converted into bacterial protein, provided that there is sufficient readily fermentable energy available to the bacteria. However, any ammonia which is not used by the microbes will be absorbed from the rumen, transformed into urea in the liver, and passed into the blood; the urea in the blood will be excreted in the urine and the milk. Thus it can be predicted theoretically that the concentration of urea in milk should increase if there is a surplus of rumen-degradable protein in the diet and/or if there is a deficiency of energy in the diet, and that the concentration should decrease if there is a deficiency of protein in the diet.

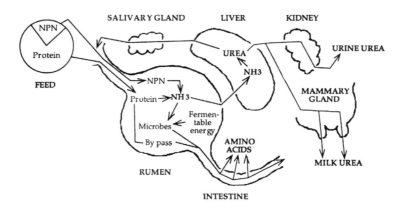

**Figure 7.6** Nitrogen metabolism in ruminants. NPN = non protein nitrogen. NH₃ = ammonia (Carlsson, 1994)

The results of recent research seem to support the practical relevance of the theoretical model presented in Figure 7.6 . Gustafsson and Palmquist (1993) observed a positive correlation between the concentration of ammonia in the ruminal fluid and the concentration of urea in blood, and several authors have reported a strong correlation between the concentrations of urea in the blood and in the milk (e.g. Oltner and Wiktorsson, 1983). It has also been shown that a surplus of crude protein in the diet gives rise to a high concentration of urea in the blood and milk (e.g. Refsdal, Baevre and Bruflot ,1985; Ferguson, Blanchard, Galligan, Hoshall and Chalupa, 1988), and Carlsson and Pehrson (1993) observed very low milk urea concentrations in samples taken from cows on farms where the rations were low in protein. Furthermore, it has been demonstrated that the concentration of urea in blood and milk is affected not only by the dietary intake of digestible crude protein, but also by the balance between the quantities of energy and protein in the diet (e.g. Oltner and Wiktorsson, 1983; Carlsson, 1994; Figure 7.7).

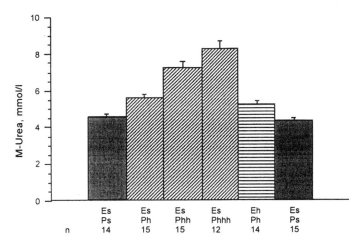

**Figure 7.7** Mean milk urea concentration (mmol/l) in cows fed different amounts of metabolisable energy (ME) and digestible crude protein (DCP). EsPs = standard feeding of ME and DCP; Ph = 300 g excess DCP; Phh = 600 g excess DCP; Phhh = 900 g excess DCP; EhPh = 25 MJ excess ME and 300 g excess DCP. Bars = s.e. n = number of samples (Carlsson, 1994)

Towards the end of the 1980s a new system for the evaluation of dietary protein was introduced in the Nordic countries (Madsen, 1985). It takes into account not only the balance between the amounts of rumen-digestible protein and fermentable energy available for the synthesis of microbial protein in the rumen (PBV), but also the quantities of amino acids absorbed from the small intestine (AAT). In the light of this new system, the results of a study by Carlsson and Pehrson (1994) further illustrate the value of measurements of milk urea concentration; as expected there was a significant correlation between the milk urea concentration and PBV, but none between milk urea and AAT.

Carlsson and Pehrson (1994) concluded that when cows are fed typical Swedish rations and are milked twice a day a milk urea concentration between 4.0 and 5.5 mmol/l should indicate that their diet is balanced with respect to energy and protein. (If cows are fed other basic feedstuffs and managed in different ways then a well balanced diet might be indicated by a different range of urea concentrations, and a new "normal" range would have to be established). A similar range of normal milk urea concentrations was observed by Gustafsson and Carlsson (1993), and they also concluded that optimal fertility was achieved when the milk urea concentration was between 4.5 and 5.0 mmol/l (Figure 7.8), their results were consistent with earlier reports that the fertility of cows was reduced when their milk urea concentration was either high (e.g. Carroll, Barton, Anderson and Smith,

1988; Ferguson *et al.* , 1988; Canfield, Sniffen and Butler, 1990; Pehrson *et al.*, 1992) or low (Pehrson *et al.* , 1992; Carlsson and Pehrson, 1993).

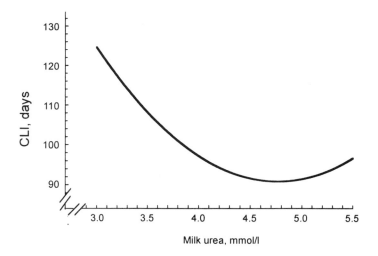

**Figure 7.8** The relationship between urea concentration in milk (mmol/l) and calving to last service interval (CLI, days). (From Gustafsson and Carlsson, 1993)

Refsdal (1983) and Carlsson, Bergström and Pehrson (1995) found that the concentration of urea in bulk milk can be used as a reliable guide to the nutritional efficiency of the diet of a herd.

There is increasing interest internationally in using milk urea as a biological indicator of the efficiency of the diets fed to dairy cows. However, until more is known about the factors which affect milk urea concentration, the results must be interpreted with caution and only after taking into consideration the variations in feeding practices and management systems in different countries. It is necessary to establish the effect on milk urea of the catabolism of tissue proteins and of the amino acids absorbed from the digestive tract. It is also necessary to be aware that there is a considerable daily variation in milk urea concentration and that milk with a very high fat content may give misleading results (Carlsson and Bergström, 1994). There are also variations with the stage of lactation, particularly during the first few weeks when the milk urea concentration is lower than later in lactation (Gustafsson, 1993).

## Allantoin

The synthesis of protein by the microbes in the rumen is of vital importance for the health and productivity of ruminant animals. The conventional methods for measuring the rate of synthesis of this protein require the use of animals fitted with abomasal or duodenal cannulae; these methods are not only laborious and expensive but also questionable from the point of view of the animals' welfare. During the last decade promising results have been obtained by measuring the concentrations of derivatives of purine (allantoin, uric acid, xanthine and hypoxanthine), which are the products of the breakdown of the nucleic acids DNA and RNA, as indicators of the supply of microbial protein to ruminants.

Nucleic acids are derived directly from the diet, from the rumen microbes and from endogenous catabolism. However, the bulk of the purine derivatives which are excreted is derived from the rumen microbes (Antoniewicz and Pisulewski, 1982; Giesecke, Ehrentreich, Stangassinger and Ahrens, 1994), and the other two sources contribute only a little to the total excretion. As a result the metabolic conditions are favourable for using measurements of the excretion of purine derivatives as an indicator of the rate of production of microbial protein in the rumen. Allantoin, which is the final breakdown product of nucleic acids, has been regarded as the most interesting of the purine derivatives, mainly because it appears to be the main excretory product and is excreted in the urine and the milk.

More than 90% of the allantoin excreted is excreted in the urine, and most trials have used measurements of urinary allantoin for evaluating the production of protein in the rumen. However, in spite of the fact that only 1 to 4% of total allantoin excreted appears in the milk (Giesecke *et al.,*, 1994; Gonda, 1995), several authors have preferred to use milk as the test medium.

The concentration of allantoin in milk, measured by high performance liquid chromatography, has been reported to range from about 100 to 600 µmol/l (Rosskopf and Giesecke, 1992; Giesecke *et al.* , 1994; Gonda 1995). The results have indicated that there is a non-linear relationship between milk yield and allantoin excretion, and close correlations between the dietary energy intake (Rosskopf and Giesecke, 1992) and dry matter intake (Kirchgessner and Kreuzer, 1985; Gonda, 1995) and the concentration of allantoin in milk. Giesecke *et al.* (1994) concluded that the measurement of milk allantoin appears to be a useful non-invasive method for monitoring the synthesis of rumen microbial protein; however, Gonda (1995) doubted the relevance of this conclusion because of the high correlation between the amount of allantoin excreted in milk and the milk yield.

It is evident that further research will be necessary before the value of milk allantoin as an indicator of the synthesis of microbial protein can be assessed adequately.

## Vitamins and trace elements

The scientific information in this field is very restricted and further research is urgently required. However, it seems reasonable to assume that the concentration in milk of at least some of the vitamins and trace elements may be related to their nutritional balance. In the case of selenium it has been found experimentally that its concentration in milk is well correlated with the content of selenium in the diet (Aspila, 1991), and in Sweden practical experience has been in accordance with Aspila's results. Sweden is a selenium deficient country and before 1980 the average concentration of selenium in milk supplied to consumers was about 7 µg/l; in that year it was decided to supplement all mineral additives for farm animal feedstuffs with 10 mg selenium/kg; the concentration of selenium in milk then increased to about 11 µg/l. Later the level of supplementation was increased to 30 mg/kg and the average concentration of selenium in milk is now about 15 µg/l (Pehrson, 1993).

It is expensive to analyse samples directly for selenium. However, it is much easier and cheaper to measure the activity of the selenium-containing enzyme glutathione peroxidase (GSH-Px) and this enzyme is often used instead of selenium for estimating the selenium status of blood. An ongoing project in our laboratory is aimed to evaluate the correlation between selenium concentration and GSH-Px activity also in milk.

## Final conclusions

The cost of feedstuffs constitutes the largest part of the total cost of dairy production. As the demand for overall economic efficiency increases it becomes steadily more important that the efficiency of feeding a dairy herd should be maintained at an optimal level, to try to improve both the productivity and the health of the herd. This paper has shown that several of the constituents of milk which can be analysed at a reasonable cost are useful as aids to achieving this aim. The interest in the use of milk profiles is increasing, and there is good reason to believe that the information derived from the parameters so far included should become steadily more reliable. It may also be expected that new, practically useful parameters will be included in the milk profile, to the benefit of both the health of dairy herds and the profit of farmers.

## References

Adams, R.S., Stout, W.L., Kradel, D.C., Guss, S.B. Jr., Mosel, B.L. & Jung, G.A. (1978) Use and limitations of profiles in assessing health or nutritional status of dairy herds. *Journal of Dairy Science*, **61**, 1671–1679

Ahlin, K.Å. and Larsson, K. (1985) Variation i mjölkprogesteron vid igångsättning av cyklicitet i äggstockarna hos mjölkkor. (Variations in milk progesterone levels at the onset of cyclicity of the ovaries in dairy cows). *Proceedings NKJ*, Mankans, Finland, pp 345–353

Andersson, L. (1984) Detection, occurrence, causes and effects of hyperketonaemia in Swedish dairy cows. *Thesis*. Swedish University of Agricultural Sciences, Skara, Sweden, 80pp

Andersson, L. (1988) Subclinical ketosis in dairy cows. In *The Veterinary Clinics of North America: Food Animal Practice*, **4**, pp 233–251. Edited by T.H. Herdt. W.B. Saunders Company, Philadelphia, PA, USA

Andersson, L. and Emanuelson, U. (1985) An epidemiological study of hyperketonaemia in Swedish dairy cows: determinants and the relation to fertility. *Preventive Veterinary Medicine*, **3**, 449–462

Andersson, L. and Lundström, K. (1985) Effect of feeding silage with high butyric acid content on ketone body formation and milk yield in postparturient dairy cows. *Zentralblatt für Veterinärmedizin A*, **32**, 15–23

Antoniewicz, A.M. and Pisulewski, P.M. (1982) Measurement of endogenous allantoin excretion in sheep urine. *Journal of Agricultural Science, Cambridge*, **98**, 221–223.

Aspila, P. (1991) Metabolism of selenite, selenomethionine and feed-incorporated selenium in lactating goats and dairy cows. *Journal of Agricultural Science in Finland*, **63**, 1–74

Berglund, B. and Larsson, K. (1983) Milk ketone–bodies and reproductive performance in post partum dairy cows. *Proceedings Fifth International Conference on Production Disease in Farm Animals*, Uppsala, Sweden, pp 153–157

Blowey, R.W. (1975) A practical application of metabolic profiles. *Veterinary Record*, **97**, 324–327

Canfield, R.W., Sniffen, C.J. and Butler, W.R. (1990) Effects of excess degradable protein on postpartum reproduction and energy balance in dairy cattle. *Journal of Dairy Science*, **73**, 2342–2349

Carlsson, J. (1994) The value of the concentration of urea in milk as an indicator of the nutritional value of diets for dairy cows, and its relationships with milk production and fertility. *Thesis*. Swedish University of Agricultural Sciences, Uppsala, Sweden

Carlsson, J. and Bergström, J. (1994) The diurnal variation of urea in cow's milk and how milk fat content, storage and preservation affects analysis by a flow injection technique. *Acta veterinaria scandinavica*, **35**, 67–77

Carlsson, J., Bergström, J. and Pehrson, B. (1995) Variations with breed, age, season, yield, stage of lactation and herd in the concentration of urea in bulk milk and in individual cow's milk. *Acta veterinaria scandinavica*, **36**, 245–254

Carlsson, J. and Pehrson, B. (1993) The relationships between seasonal variations in the concentration of urea in bulk milk and the production and fertility of dairy herds. *Journal of Veterinary Medicine A*, **40**, 205–212

Carlsson, J. and Pehrson, B. (1994) The influence of the dietary balance between energy and protein on milk urea concentration. Experimental trials assessed by two different protein evaluation systems. *Acta veterinaria scandinavica*, **35**, 193–205

Carroll, D.J., Barton, B.A., Anderson, G.W. and Smith, R.D. (1988) Influence of protein intake and feeding strategy on reproductive performance of dairy cows. *Journal of Dairy Science*, **71**, 3470-3481

Dirksen, G. (1994) Kontrolle von Stoffwechselstörungen bei Milchkühen an Hand von Milchparametern. (Control of metabolic disturbances in dairy cows by milk parameters). Proceedings of XVIII World Buiatrics Congress, Bologna, Italy, pp 35–45

Dohoo, I.R. and Martin, S.W. (1984) Subclinical ketosis: prevalence and associations with production and disease. *Canadian Journal of Comparative Medicine*, **48**, 1-5

Emery, R.S. (1978) Feeding for increased milk protein. *Journal of Dairy Science*, **61**, 825–828

Engvall, A. (1980) Low milk fat syndrome in Swedish dairy cows. Field and experimental studies with special reference to the rumen microbiota. *Acta veterinaria scandinavica*, Suppl. 72, 124pp

Ferguson, J.D., Blanchard, T., Galligan, D.T., Hoshall. D.C. and Chalupa, W. (1988) Infertility in dairy cattle fed a high percentage of protein degradable in the rumen. *Journal of the American Veterinary Medical Association*, **192**, 659–662

Foster, L.A. (1988) Clinical ketosis. In *The Veterinary Clinics of North America: Food Animal Practice*, 4, pp 253–267. Edited by T.H. Herdt. W.B. Saunders Company, Philadelphia, PA, USA

Gonda, H.L. (1995) Nutritional status of ruminants determined from excretion and concentration of metabolites in body fluids. *Thesis*. Swedish University of Agricultural Sciences, Uppsala, Sweden

Gordon, F.J. (1977) The effect of protein content on the response of lactating cows to level of concentrate feeding. *Animal Production*, **25**, 181–191

Gustafsson, A.H. (1993) Acetone and urea concentration in milk as indicators of the nutritional status and the composition of the diet of dairy cows. *Thesis*. Swedish University of Agricultural Sciences, Uppsala, Sweden, 143pp

Gustafsson, A.H., Andersson, L. and Emanuelson, U. (1993) Effect of hyperketonaemia, feeding frequency and intake of concentrate and energy on milk yield in dairy cows. *Animal Production*, **56**, 51–60

Gustafsson, A.H. and Carlsson, J. (1993) Effects of silage quality, protein evaluation systems and milk urea content on milk yield and reproduction in dairy cows. *Livestock Production Science*, **37**, 91–105

Gustafsson, A.H. and Palmquist, D.L. (1993) Diurnal variation of rurmen ammonia, serum urea, and milk urea in dairy cows at high and low yield. *Journal of Dairy Science*, **76**, 475–484

Giesecke, D., Ehrentreich, L., Stangassinger, M. and Ahrens, F. (1994) Mammary and renal excretion of purine metabolites in relation to energy intake and milk yield in dairy cows. *Journal of Dairy Science*, **77**, 2376–2381

Hagert, C. (1991) Kontinuierliche Kontrolle der Energie– und Eiweissversorgung der Milchkuh während der Hochlaktation an Hand der Konzentrationen von Azeton, Harnstoff, Eiweiss und Fett in der Milch. (Continous control of the energy and protein balance in dairy cows during peak lactation concerning acetone, urea, protein and fat in milk). *Thesis*, University of Munich, Germany

Henkel, H. and Borstel, E. v. (1969) Untersuchungen über die Abhängigkeit des Auftretens der Azetonämie vom Ernährungshaushalt hochleistender Milchkühe. (Studies on the interrelationship between the feed economy of high–yielding dairy cows and the occurrence of acetonaemia.). *Archiv für Tiernährung*, **19**, 259–272

Herdt, T.H. (1988) Fuel homeostasis in the ruminant. In *The Veterinary Clinics of North America: Food Animal Practice*, **4**, pp 213–231. Edited by T.H. Herdt, W.B. Saunders Company, Philadelphia, PA, USA

Kirchgessner, M. and Kreuzer, M. (1985) Harnstoff und Allantoin in der Milch von Kühen während und nach Verfütterung zu hoher und zu niedriges Proteinmengen. (Urea and allantoin in milk from dairy cows during and after overfeeding and underfeeding with protein). *Zeitschrift für Tierphysiologie, Tierernährung und Futtermittelkunde*, **54**, 141–151

Lucey, S., Rowlands, G.J., and Russel, A.M. (1986) Short term associations between disease and milk yield of dairy cows. *Journal of Dairy Research*, **53**, 7-15

Madsen, J. (1985) The basis for the proposed Nordic protein evaluation system for ruminants. The AAT–PBV system. *Acta Agriculturae Scandinavica*, Suppl. 25, pp 9–20

Oltner, R. and Wiktorsson, H. (1983) Urea concentrations in milk and blood as influenced by feeding varying amounts of protein and energy to dairy cows. *Livestock Production Science*, **10**, 457–467

Øverby, I., Aas Hansen, M., Jonsgård, K. and Søgnen, E. (1974) Bovine ketosis. I. Occurrence and incidence in herds affected by ketosis in eastern Norway 1967–1968. *Nordisk Veterinärmedicin*, 26, 353–361

Payne, J.M., Dew, S.M., Manston, R. & Faulks, M. (1970) The use of a metabolic profile test in dairy herds. *Veterinary Record*, **87**, 150–158

Pehrson, B. (1966) Studies on ketosis in dairy cows. Thesis, the Royal Veterinary College, Stockholm, Sweden. *Acta veterinaria scandinavica*, Suppl. 15, 59pp

Pehrson, B. (1993) Selenium in nutrition with special reference to the biopotency of organic and inorganic compounds. In *Biotechnology in the Feed Industry. Proceedings of Alltech's Ninth Annual Symposium.* 71–89. Edited by T.P. Lyons. Alltech Technical Publications, Nicholasville K.Y., USA

Pehrson, B., Plym Forshell, K. and Carlsson, J. (1992) The effect of additional feeding on the fertility of high–yielding dairy cows. *Journal of Veterinary Medicine A*, **39**, 187–192

Plym Forshell, K., Andersson, L and Pehrson, B. (1991) The relationships between the fertility of dairy cows and clinical and biochemical measurements, with special reference to plasma glucose and milk acetone. *Journal of Veterinary Medicine A*, **38**, 608–616

Refsdal, A.O. (1982) Ovariecyster hos melkekyr. (Ovarial cysts in dairy cows). *Norsk Veterinär Tidsskrift*, **94**, 789–796

Refsdal, A.O. (1983) Urea in bulk milk as compared to the herd mean of urea in blood. *Acta veterinaria scandinavica*, **24**, 518–520

Refsdal, A.O., Baevre, L. and Bruflot, R. (1985) Urea concentration in bulk milk as an indicator of the protein supply at the herd level. *Acta veterinaria scandinavica*, **26**, 153–163

Rosskopf, R. and Giesecke, D. (1992) Untersuchungen an Kühen über den Einfluss der Energieaufnahme auf den Pansenstoffwechsel mittels der Allantoinausscheidung in der Milch. (Investigations in cows on the influence of energy intake on rumen metabolism by means of allantoin excretion in the milk). *Journal of Veterinary Medicine A*, **39**, 515–524

Saloniemi, H. (1995) Use of somatic cell count in udder health work. In *The Bovine Udder and Mastitis*, pp 105–110. Edited by M. Sandholm, T. Honkanen-Buzalski, L. Kaartinen and S. Pyörälä. Jyväskylä, Finland, Gummerus Kirjapaino Oy

Simensen, E., Halse, K., Gillund, P. and Lutnaes, B. (1990) Ketosis treatment and milk yield in dairy cows related to milk acetoacetate levels. *Acta veterinaria scandinavica*, **31**, 433–440

Spörndly, E. (1989) Effects of diet on milk composition and yield of dairy cows with special emphasis on milk protein content. *Swedish Journal of Agricultural Research*, **19**, 99–106

Storry, J.E. (1970) Reviews of the progress of dairy science. Section A. Physiology. Ruminant metabolism in relation to the synthesis and secretion of milk fat. *Journal of Dairy Research*, **37**, 139–164

Sutton, J.D. (1989) Altering milk composition by feeding. *Journal of Dairy Science*, **72**, 2801–2814

Taponen, J. and Myllys, V. (1995) The economic impact of mastitis. In *The Bovine Udder and Mastitis*, pp 261–264. Edited by M. Sandholm, T. Honkanen—Buzalski, L. Kaartinen and S. Pyörälä. Jyväskylä, Finland, Gummerus Kirjapaino Oy

Wiktorsson, H. (1971) Input/Output Relationships in Dairy Cows. The effects of different levels of nutrition, quantities of roughage and frequencies of feeding. *Thesis*, the Agricultural College, Uppsala, Sweden, 114pp

**8**

# THE IMPORTANCE OF GRASS AVAILABILITY FOR THE HIGH GENETIC MERIT DAIRY COW

D.A. MCGILLOWAY and C.S. MAYNE
*Agricultural Research Institute of Northern Ireland, Hillsborough, Co. Down, BT26 6DR.*

## Introduction

The introduction of milk quotas in 1984 effectively set ceiling limits to milk production and provided for a commodity price well above world market prices. For the individual farmer this has resulted in increased emphasis on efficiency of production per litre, rather than production *per se*. Efficiency of production can be increased either by reducing inputs whilst maintaining output, or by increasing both inputs and outputs such that the value of the additional output exceeds the cost of the extra input. Adoption of these strategies has resulted in a gradual shift in emphasis on many dairy farms towards maximising milk output per cow, at the expense of output per hectare. A logical extension of this policy is to breed the highest yielding cows to the best available bulls. This has been achieved with considerable success in the USA, Canada and Holland, etc., where the chosen system of production relies heavily on complete diet feeding and intensive feedlot management. The question is - do cows that are genetically predisposed to produce more milk under high input systems have a role to play in low input grass-based systems of production?

The idea that grass might have a role to play in feeding cows of high genetic merit is not a prospect that feed manufacturers are likely to welcome with open arms. However, the reality is that if high merit cows are to make a useful contribution to the farming economies of countries which operate low cost systems of milk production based on grass and grass-silage, then they must do so whilst continuing to utilize pasture grazed *in situ* as the greater portion of the diet. Scope exists for modifying existing management systems to accommodate the high merit cow, but in low input

135

systems, there are fewer alternatives for change than in more complex production systems that use a range of inputs, produced either on the farm, or purchased from outside sources. Innovative grassland management strategies, coupled with supplementation of grazed grass with complementary concentrates of high nutrient concentration, offers the best hope for resolving this dilemma in low input systems, provided it can be achieved without large-scale substitution of grass. Given the suitability of climatic conditions in western regions of the United Kingdom (UK) and Ireland for growing high quality grass at minimal cost, and the more favourable cereal and forage maize growing conditions prevailing elsewhere in Europe, such an approach affords the best opportunity to maintain the competitiveness of low input systems. If this objective cannot be realised, then continued selection for increased genetic merit in grass-based milk production systems must be questioned.

It is also worth noting that given the imposition of GATT and other reforms, milk price in Europe will increasingly reflect the market price for the product. Consequently, the long term viability of systems currently dependent upon high milk prices must be called into question.

Can we feed the high merit cow on a predominantly grass-diet and still achieve satisfactory levels of performance? What are the limitations to intake, and how might they be overcome? These are the issues that the authors have sought to address in this review. Also included are definitions of the principal terms of reference, i.e. herbage availability and high genetic merit.

## Definitions

### HERBAGE AVAILABILITY

For the purposes of this review, herbage availability is defined as the relative ease or difficulty with which herbage can be harvested by the grazing animal (Wade, 1991). This is a qualitative term and is considered to be distinct from herbage allowance which is a quantitative measure of the herbage on offer, but does not take account of the manner in which herbage is presented to the grazing animal. Herbage availability therefore represents a description of the interaction between the sward and the grazing animal. Sward characteristics known to influence the grazing process include canopy height, leaf length and extended tiller height (Wade, 1991), herbage bulk density (Laca, Ungar, Seligman and Demment, 1992), stem and pseudostem position in the canopy (Flores, Laca, Griggs and Demment, 1993), and leaf orientation, stiffness and tensile strength (Laca, Demment, Distel and Griggs, 1993). Nutritional factors such as protein and carbohydrate fractions must also be considered.

## HIGH GENETIC MERIT

The term 'high genetic merit' is used to describe a cow, which as the result of selection for yield traits, is genetically predisposed to produce significantly more milk and milk constituents than a cow of lower merit status. The term is relative: what was considered as being high merit in the mid 80's, is by today's standards more accurately described as of intermediate status. In Ireland and the UK, the rate of genetic progress for yield traits has increased considerably since the mid 1980's. For example the rate of genetic progress in sires for fat plus protein yield has trebled in the ten year period between 1980 and 1990 (Figure 8.1) (UK Statistics for Genetic Evaluation, July 1995).

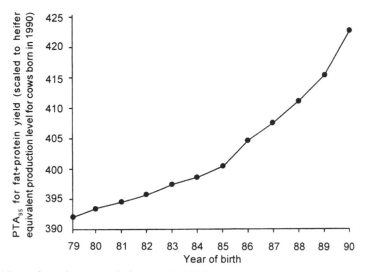

**Figure 8.1** Rate of genetic progress in fat+protein yield for sires born over the 12 year period 1979-1990. Source: UK Statistics for Genetic Evaluation (July 1995)

The term Predicted Transmitting Ability (PTA $_{95}$) is a measure of an animals ability to transmit its genes to the next generation, and can relate to production traits, compositional quality, linear type, etc. It is measured relative to a fixed reference point, the genetic base, which is the average PTA of animals born during 1990 and milking in 1995. However, the selection of animals using PTAs for various traits, e.g. fat+protein yield, has now been largely superseded by selection based on indices such as the Profit Index (PIN) and more recently the Index of Total Economic Merit (ITEM). These are overall breeding indices which weight PTAs for various economically important traits, after accounting for the costs of milk production, into an overall financial value. This value therefore indicates the predicted margin over food and quota costs that the animal is expected to pass on to its progeny.  A similar index, the Relative Breeding Index (RBI) is used in Ireland, and takes account of the Irish dairy situation to weight PTAs into an overall breeding index.  In essence therefore, indices such as PIN, ITEM and RBI enable dairy farmers to select animals that will improve farm profits.  Given that many dairy farmers are breeding to a clearly defined goal, selection based on these indices can be expected to further increase the rate of genetic progress.  The consequence of this is that although the PIN value of the UK national dairy herd is currently around £17, some breeders have herds with PIN values in excess of £38 (M.Coffey, *pers comm.*).  If this trend continues, then by the year 2000 there will be numerous herds with average PIN values greater than £50, and individual animals with PIN values in excess of £100 (A.Cromie, *pers comm.*).

## Implications of increased genetic merit for pasture-based production systems

Before considering the importance of herbage availability for the high merit cow, it is worthwhile considering the likely implications of increased genetic merit for pasture based production systems in general.

GENOTYPE X ENVIRONMENT INTERACTIONS

There has been much discussion recently on the possible existence of an interaction between various environmental factors (principally nutrition) and genotype.  The existence of such an interaction would have major implications for the dairy industry, since it would mean that an animal performing well in one system of production, e.g. intensive feedlot, might not necessarily be able to hold that production advantage in a low input system.  The problem then for producers in opting to select from American and continental bred sires – sires that have been selected and assessed on the basis of daughter performance on complete diets fed indoors, with limited access to grazed grass – is that there is no indication of how well a particular bull will perform in a grass-based system.  Considerable time and money could be invested in a selection *cul de sac*.

Evidence of a genotype x environment interaction can be found in the results of a joint New Zealand (NZ) and Canadian study (CANZ) in which the daughters of Canadian and NZ sires were reared and milked in Canada or NZ (Peterson, 1988; Burnside, Graham, Rapitta, McBride and Gibson, 1991; Graham, Burnside, Gibson, Rapitta and McBride, 1991).  Overall, progeny of NZ sires produced fractionally less milk and milk solids than the Canadian daughters in both countries, though differences between sire groups were greater when evaluated in Canadian herds, i.e. NZ daughters produced more milk fat in Canada than in NZ, due primarily to the more generous feeding in Canada (heifers weighed 530 kg in Canada compared with 390 kg in NZ) (Table 8.1). Further analysis of the data revealed considerable differences in the ranking of Canadian sires based on the performance of their daughters in NZ, compared to that predicted from evaluation in Canada (Table 8.2). This suggests that sire proofs for milk, fat and protein made under Canadian feeding and management conditions are not good predictors of a bull's relative merit in the NZ environment. According to Peterson (1988), the fact that the Canadian sires ranked differently in the NZ environment implies that the trait limiting production in the two countries is different. Given the almost total dependence on grazed grass in NZ, the results suggest that daughters of some Canadian sires performed significantly better on a pasture-based diet (when compared to daughters of other Canadian sires) than would be predicted from their performance in Canada, i.e. on a diet with a high nutrient concentration. This could reflect differences in grazing behavioural characteristics between daughters of different sires. Choice of appropriate sires is therefore of considerable importance in situations where reliance on grazed grass is high. This subject has been discussed in further detail by Holmes (1995), and Mayne and Gordon (1995).

**Table 8.1** PERFORMANCE OF PROGENY OF CANADIAN OR NEW ZEALAND SIRES EVALUATED EITHER IN CANADA OR NEW ZEALAND.

| Trait | Location (for evaluation) | Sire Group Canadian | New Zealand |
|---|---|---|---|
| Milk (kg) | Canada | 6097 | 5469 |
| | New Zealand | 3395 | 3157 |
| Fat (kg) | Canada | 231 | 226 |
| | New Zealand | 140 | 138 |
| Protein (kg) | Canada | 206 | 192 |
| | New Zealand | 109 | 105 |
| Fat + Protein (kg) | Canada | 437 | 418 |
| | New Zealand | 249 | 243 |

Source : Peterson (1988); Burnside, Graham, Rapitta, McBride and Gibson (1991).

**Table 8.2** CORRELATION BETWEEN SIRE RANKINGS IN CANADIAN/NEW ZEALAND
TRIAL WITH THOSE IN COUNTRY OF ORIGIN.

| Trait | Canadian sires | New Zealand sires |
|---|---|---|
| Milk (kg) | 0.22[+] | 0.80 |
| Fat (kg) | 0.25[+] | 0.79 |
| Protein (kg) | 0.36[+] | 0.80 |
| Fat (g/kg) | 0.69 | 0.75 |
| Protein (g/kg) | 0.68 | 0.54 |
| Expected correlation | 0.68 | 0.65 |

[+]Values significantly different to expected values

Source : Peterson (1988).

Further evidence of a genotype x nutrition interaction was presented by
Veerkamp, Simm and Oldham (1994), who undertook a regression analysis (based
on Langhill data from Scotland) of animal performance against pedigree index
(PI), and tested regression coefficients for possible PI x diet (proportion of forage)
interactions. Significantly different regression coefficients were observed between
PI and milk yield for the low and high forage diets respectively, indicating that
differences between merit groupings vary with the forage content of the diet.

Given the existence of genotype x nutrition interactions, it is possible that high
genetic merit heifers and cows (bred to perform on diets of high nutrient
concentration fed indoors), such as those imported into Hillsborough (N. Ireland)
and Moorepark (Ireland), will not be as well suited to a grass-based diet as animals
evaluated under grass-based production systems. Consequently, the response of
these animals in a grazing situation is likely to underestimate the true potential of
high merit cows that have been selected on the basis of intake and foraging
characteristics.

Possible genotype x nutrition interactions on fertility and animal health have
not been studied, though it is likely that they exist and are of some importance
(Holmes, 1995). Cows producing higher volumes of milk are under more
physiological stress than lower producing cows (Funk, 1993). Several studies
(Shanks, Freeman, Berger and Kelley, 1978, and Bertrand, Berger, Freeman and
Kelley, 1985) have shown that health costs tend to increase as production increases.
Hansen, Freeman and Berger (1983) and Oltenacu, Frick and Lindhe (1991) have
shown a negative genetic correlation between cow reproductive performance and
production, suggesting that continual selection for production will result in poorer

reproductive performance. In a grazing trial at Moorepark with cows of low to intermediate merit status, grazing to residual sward heights of 40, 60 and 80 mm resulted in an inverse linear relationship between residual sward height and calving to pregnancy interval (Ryan, Snijders, McGilloway and O'Farrell, *unpublished*). It can be expected that high merit cows in the same situation would be seriously disadvantaged. Veerkamp *et al.*, (1994) make the point that because tissue reserves in dairy cows are substantial (Butler-Hogg, Wood and Bines, 1985; Gibb, Ivings, Dhanoa and Sutton, 1992), high merit cows might use these reserves to buffer against nutritional adversity in the short term, with interactions only becoming evident in the longer term (i.e. in subsequent lactations).

## FOOD INTAKE

Given the high merit dairy cow as previously defined, there is a lack of data with regard to food intake at pasture. Consequently, it is necessary to look to the indoor environment to gain an insight into intake potential and feed conversion efficiency.

It is now generally accepted that improvements in milk production with increasing genetic merit stem from a change in nutrient partitioning within the cow, whereby a greater proportion of nutrient intake is diverted towards milk production at the expense of liveweight gain (Sejrsen and Neimann-Sorensen, 1994; Veerkamp *et al.*, 1994; Patterson, Gordon, Mayne, Porter and Unsworth, 1995). A summary of indoor feeding trials at Langhill (Scotland) and more recently at Hillsborough are presented in Table 8.3. In both studies, animals of intermediate or high merit status produced significantly more milk, and fat plus protein, than low merit animals. However, intake of energy did not differ significantly between the various genetic groupings. In the Hillsborough study, although high merit cows produced almost 22 per cent more fat plus protein per unit food consumed, intakes of dry matter (DM) were only slightly higher than with cows of intermediate genetic merit. In addition the high merit cows lost up to 1 kg liveweight per day over the first 60 days of lactation, despite being on a high level of concentrate input (14 kg concentrate DM/d).

Similar responses have been observed in grazing trials carried out in NZ. At pasture, cows of 'high' merit produced more milk (by 20-40%), consumed more herbage (by 5-20%), were more efficient converters of food into milk (by 10-15%) and produced higher milk yields per hectare than low merit cows (Grainger, Holmes and Moore, 1985; Holmes, 1988; Holmes, 1995). It is important to note that the difference in genetic merit of the high and low groups used in these NZ experiments was small compared to that in the Langhill and Hillsborough studies.

**Table 8.3**  EFFECTS OF INCREASED GENETIC MERIT ON FEED EFFICIENCY (SILAGE BASED SYSTEMS).

|  | Langhill studies (182 days) | | Hillsborough studies (160 days) |
| --- | --- | --- | --- |
| *Genetic merit (PTA$_{90}$ kg F + P)* | 4.3 | *vs* 18.8 | 5 vs 45 |
| *Comparative treatment* | *High conc.* | *Low conc.* | *High conc.* |
| **Animal performance (% change)** | | | |
| Food intake | +5.0 | +4.3 | +6.3 |
| Fat + protein yield | +11.5 | +12.2 | +29.6 |
| **Food conversion efficiency (% change)** | | | |
| Fat + protein yield/unit food intake | +6.2 | +7.6 | +21.8 |

Source : Veerkamp, Simm and Oldham (1994); Patterson, Gordon, Mayne, Porter and Unsworth (1995).

Voluntary food intake is a major factor influencing total nutrient intake and hence degree of body tissue mobilization in early lactation (Forbes, 1995). Undoubtedly the physical capacity of the digestive tract and/or the rate of substrate utilization are important aspects of intake regulation, but other factors also exert influence. For example, metabolic control is thought to be mediated via chemo- and osmo- receptors in the rumen, intestines and liver, which are sensitive to volatile fatty acids (VFAs). Higher yielding cows absorb VFAs more quickly than low yielders, principally via a greater demand from the mammary gland, resulting in weaker negative feedback from these receptors and hence increasing food intake (Gill and Beever, 1991; Forbes, 1995). Nevertheless, the available data indicate that although high merit cows have higher intake characteristics than animals of lower merit, the difference is small in comparison with the large differences observed in milk production.

*Future selection strategies*

Selection for yield traits without due consideration of food intake may well be seen as a major oversight by breeders in the years to come. According to Veerkamp, Emmans, Cromie and Simm, (1995) food intake in dairy cattle (under winter feeding regimes) has a heritability of 0.36, indicating that direct selection will result in considerable genetic improvement for this trait. However, in most situations it is not possible to measure intake directly, so other indirect traits may have to be considered, e.g. chest width, body depth and muzzle size, etc. Higher intakes might well partially offset the greater loss of live weight and condition observed with high merit cows in early lactation.

NUTRIENT REQUIREMENTS AT PASTURE

Mayne and Gordon (1995) calculated a theoretical energy balance for cows of medium and high merit (assuming yields of 25 and 32.5 kg milk/day respectively in early lactation) at pasture (Table 8.4).Assuming similar live-weight loss, herbage intakes of 15.0 and 18.7 kg DM/cow/d were predicted to meet the energy requirements for medium and high merit cows respectively. Work by McGilloway and Stakelum (*unpublished*) at Moorepark indicates that where cows were grazed exclusively on grass, to residual sward heights of 40, 60 and 80 mm, intake of DM up until mid July averaged 13.8, 14.3 and 15.8 kg/cow/day respectively (as measured using the n-alkane technique). Previous studies in the Netherlands (Meijs and Hoekstra, 1984) indicate that 16.9 kg DM/cow/day is an upper limit to the intake of grass under 'ideal' grazing conditions. Consequently, under good grassland management practices, herbage intake at pasture is normally sufficient to meet the requirements of the medium merit cow.

**Table 8.4** THEORETICAL ENERGY BALANCE FOR MEDIUM AND HIGH GENETIC MERIT COWS ON GRAZED PASTURE.

| *Genetic merit* *(Predicted Transmitting Ability (kg Fat + Protein))* | *Medium* *(+5)* | *High* *(+60)* |
|---|---|---|
| Live-weight (kg) | 550 | 650 |
| Live-weight loss (kg/day) | 0.5 | 0.5 |
| Milk yield[+] (kg/day) | 25.0 | 32.5 |
| Metabolisable energy requirements (MJ/day) | | |
| Maintenance[++] | 55.4 | 62.5 |
| Total requirement | 180 | 225 |
| Dry matter required (kg/day) assuming ME of 12.0 MJ/kg DM) | 15.0 | 18.7 |

[+]    Assuming milk composition of 39.4 g/kg butterfat, 31.9 g/kg protein and 44.2 g/kg lactose

[++]    Maintenance energy requirements as estimated by Agricultural and Food Research Council (1993).

Source: Mayne and Gordon (1995).

It is now generally accepted that milk yield drives intake demand. Indirect evidence in support of this hypothesis can be found in the milk production response of cows treated with the hormone bST. In trials in the USA, milk production increased within days of treatment initiation, but DM intake did not respond until after four to six weeks had elapsed (Bauman, Eppard, DeGeeter and Lanza, 1985). The slower increase in intake relative to milk yield post-calving also lends support to this hypothesis. Data from grazing studies at Moorepark suggest that for each kg increase in milk yield (over the range 15-30 kg milk/day) cows will consume an extra 0.4-0.5 kg DM/day (Stakelum, 1993; Stakelum and Dillon, 1995). This relationship is depicted graphically in Figure 8.2. However for cows yielding in excess of 30 kg milk/d the nature of this relationship is unknown, but it is speculated that because of sward and animal behaviour constraints that restrict intake, the slope of the line will tend towards a plateau. The challenge in managing the high merit cow at grass, is to seek to arrest this decline in intake at high levels of animal performance, by manipulating grassland management and supplementation strategies. This implies presenting the grazing animal with herbage in a form that will maximise intake/bite and hence daily DM intake. It is accepted however, that the provision of complementary concentrates with high nutrient concentration will be a necessary component of any grazing strategy developed with high genetic merit dairy cows in early lactation. It is also important to highlight the fact that the current research work being undertaken at Hillsborough and Moorepark is unique within the UK and Ireland, with regard to feeding the high merit cow in a predominantly grass-based system

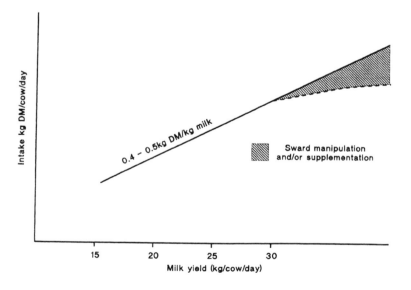

**Figure 8.2** Relationship between milk yield (kg/cow/day) and dry matter intake (kg/cow/day)

## Constraints to food intake at pasture

Limitations to the intake of herbage by the grazing animal exist at many levels. At the sward/animal interface, herbage availability and quality are the main factors constraining herbage intake. These factors are largely under the control of the farmer, but ultimately, climate and soil fertility factors exert far more powerful influences on the production system than any management strategies imposed by the farmer.

## MACRO-FACTORS

### *Climate and location*

Climatic factors largely determine the start and duration of the growing and grazing seasons, how much grass is produced, how this production is distributed across the season, and how it varies between locations and years (Brereton, 1995). Within Europe, the grass growing season can vary from 365 days in lowland coastal areas, to less than 200 days in inland continental and upland areas (Thran and Broekhuizen, 1965). Even within relatively small areas there is considerable variation. For example, Connaughton (1973) and Brereton (1995) examined the geographical variation in the start and length of the growing season in Ireland. In southern coastal areas the date of start of growth is February 1 and the length of the growing season is up to 340 days. This contrasts with inland northern areas where the equivalent values are March 20 and 240 days respectively.

Under optimum conditions, potential production from grass swards in the UK and Ireland is high, varying from 15 tDM/ha in the extreme southwest of Ireland to less than 11 tDM/ha in the northeast (Brereton, 1995). Estimates of total annual production however obscure some important effects of climate on the seasonal distribution of DM. Where the total annual production is concentrated over a short growing season, the system of herbage utilization must place relatively greater emphasis on forage production during the growing season for winter feeding. Similarly where summer drought occurs the bulk of annual production tends to be concentrated in the spring period (Brereton, Danielov and Scott, 1995).

Climate can also have quite dramatic short term effects, both within and between days, e.g. on herbage quality, where temperature and sunlight hours can affect soluble sugar concentrations within the plant. These are known to display diurnal variation, being highest around 17:00 and lowest around 08:00 (Kingsbury, 1965; Wilkinson, Price, Russell and Jones, 1994).

In effect therefore, climatic variation means that the farmer cannot predict precisely how much grass will be produced over the season, or what kind of conditions will prevail for utilization. This is in stark contrast to the situation pertaining to indoor feeding systems where diets of known quality and quantity are fed. At pasture a 'best estimate' based on feed budgeting and a retrospective examination of cow performance is often all that is available.

### Herbage varieties

Herbage varieties are considered here because it is through the plant that the production potential of a given sward is realized (Brereton *et al.*, 1995). Any species or variety will be most productive on warm, humid, highly fertile sites, but only one or two varieties will be capable of expressing the full site potential (Brereton *et al.*, 1995). The inherent characteristics of a species or variety, e.g. date of flowering, formal growth habit (erect or prostrate), the tensile strength of the leaves, etc., largely reflect the breeding policy employed in the original selection process. These factors are largely outside the sphere of influence of the farmer, and therefore cannot be manipulated in the same way as for example herbage allowance. However, selection for reduced shear strength, increased persistency, higher DM content, etc., may be desirable management objectives, and in the long term could prove of greater benefit to increasing the intake of DM by the high merit cow, than breeding strategies designed to produce more total DM per annum.

## GRASSLAND MANAGEMENT FACTORS

Poor utilization of herbage in the conventional grazing systems of the mid 80s has often been cited as a major factor limiting efficiency of production, and is closely related to the economic performance of dairy herds (Leaver, 1983; Leaver, 1987). In this scenario, herbage utilization can be increased by producing more grass per ha and /or by cows eating a greater proportion of the herbage on offer. However, from the perspective of the high merit cow, to even approach the degree of herbage utilization achieved by low to medium merit cows would result in serious energy balance deficits. If high merit cows are to achieve their potential at grass, they must have virtually unrestricted access to high quality herbage over the lactation. The implication is that utilization per hectare by the high merit cow *per se* will be poor, and therefore a major challenge in grazing management is to develop alternative strategies to utilize residual herbage to a degree commensurate with maintaining sustainable, high quality swards over the grazing season.

*Grazing system*

Providing grass of high quality over the grazing season can only be achieved by attention to detail in grassland management. We will go one step further and say that to provide quality herbage for the high merit cow, rotational grassland management systems based on fertilizer nitrogen (N) are essential. Rotational grazing has been widely advocated in countries where pasture plays a significant role in livestock production (McMeekan and Walsh, 1963). Furthermore, Evans (1981) observed a more uniform pattern of herbage intake and increased animal performance with rotational grazing relative to continuous grazing, with February-April calving dairy cows grazed at similar stocking rates. Under rotational grazing the area of pasture is divided into paddocks, the animals spending only one or two days in each, eventually coming back to the start after approximately 15-30 days. The alternative approach (continuous grazing) involves animals remaining on the same area of pasture for protracted periods.

Unless at high stocking rates, herbage production from both systems is similar (McMeekan and Walsh, 1963: Grant, Barthram and Torvell, 1981). So what is the advantage of rotational grazing over continuous? Essentially rotational grazing facilitates identification of grass surpluses and deficits far more readily than is the case where cows are continuously grazed. For example, if in a 'normal' year it takes one day to graze out paddocks to between 60 and 70 mm residual sward height (given the normal stocking rate for the farm and optimum usage of N), then reaching this same height in 0.5 days or 1.5 days would indicate that supply is failing to meet demand in the former case, and is surplus to demand in the latter. Rotational grazing permits greater flexibility, extra paddocks can be maintained as 'buffers' allowing cows to be moved onto these paddocks if conditions deteriorate (Illius, Lowman and Hunter, 1986; Brereton, 1987). Similarly, during periods of surplus growth, paddocks can be taken out of the system via big bale silage, and then fed back again towards the end of the grazing season when grass growth rates are in decline. Furthermore, rotational grazing facilitates management practices designed to utilise high residual herbage masses following grazing by high merit cows. Flexibility is the key and it is not present to the same extent in continuous grazing systems. However, the most important characteristic of rotational grazing systems is that they facilitate presentation of herbage to the animal in an 'optimum' form for prehension – see later sections.

Within rotational grazing systems, several options are available for controlling high residual sward masses which will result from under utilization by the high merit cow. Swards could be topped in order to control stem extension and maintain quality, or alternatively a leader-follower system could be implemented to utilize residual herbage. A combination of the two is the most desirable option, since topping alone is wasteful, but if the topped grass was then eaten by either cows of lower genetic merit, heifers or other dry-stock, then utilization will be enhanced, and sward quality ensured for subsequent rotations. The Dutch system of alternating grazing with cutting might also offer scope for maintaining sward quality.

### Nitrogen fertilizer

Earlier, it was stated that fertilizer-N was an essential component of the system, implying there is no role for clover. This is not necessarily the case. Clover has many merits (both nutritionally and environmentally), and has been shown to sustain higher levels of animal performance than pure grass swards (Thomson, Beever, Haines, Cammell, Evans, Dhanoa and Austin, 1985). However, it neither fixes enough N, nor does so reliably enough over the grazing season to sustain the intensive system needed to consistently meet the demands of the high merit cow. Also, there is considerable year-to-year variation in the clover content of swards, and unpredictability with sward establishment. Stewart (1985) in a review paper states 'The authors experience in managing a low N grass-clover (cv. Blanca) sward over the last 7 years, involving integrated cuts for silage with rotational grazing by beef cattle, would indicate that it may take many years for a satisfactory stable equilibrium to be reached between the grass and clover components of the sward'.

Fertilizer N is the cornerstone of current systems of rotational grazing, producing between 15 to 25 kg DM/kg N applied (Holmes, 1968; Reid, 1978). However, under farm conditions some studies have measured a response much less than this, typically in the region of 8 kg DM/kg N applied (Ball, Molloy and Ross, 1978; Leaver, 1985). Nevertheless, within the limitations imposed by the environment, fertilizer N is the most important management input 'controlling' herbage yield (Morrison, Jackson and Williams, 1974). Furthermore, when applied strategically throughout the season it can be used to improve 'mid-season' production without reducing annual yield (Morrison, 1977; Morrison, Jackson and Sparrow, 1980) thereby producing a more uniform growth pattern through the grazing season.

### Stocking density

Stocking density defines the number of animals contained within a particular paddock at any time. Production per hectare increases with increased stocking density to a maximum, and then declines rapidly. Conversely, as stocking density is increased, production per animal declines. Journet and Demarquilly (1979) estimated that for each increment in stocking density of one cow per hectare, milk production per cow was reduced by 10%, but production per hectare was increased by over 20%.

The effects of stocking density on milk production and animal live weight are exhibited through effects on herbage allowance and therefore intake. At high stocking densities animals are forced to graze deeper in the sward, i.e. to lower residual sward heights, and therefore consume a diet of lower digestibility and nutrient value than those at lower stocking densities. Interanimal space is reduced at high stocking density, thus competition for food is potentially increased (Phillips, 1993). Increased levels of social interaction may also alter grazing patterns and subsequent production levels. Stocking density will also impinge on sward dynamics, since with less time to select, animals cannot be as selective with regard to species and/or morphological characteristics (Struth, Brown, Olson, Araujo and Alijoe, 1987).

### Grazing frequency

Under conditions of good grassland management, the dairy cow is allocated fresh grass once or twice daily. In any grazing period, as time progresses, less pasture is available due to a diminishing supply of herbage, and as a result of fouling. For the individual cow, it is to her advantage to consume as much of her daily requirement as possible within a few hours of being given a fresh allowance. Thus cows with an aggressive appetite are likely to have higher intake rates than more passive eaters, or those with a lesser capacity to consume large volumes of herbage, due to constraints such as satiety and rumen fill. In conditions where cows graze in close proximity to each other, aggressive animals will be further advantaged relative to more placid animals. However, whilst 'aggressive' traits may be desirable with regard to intake levels, ingestion of large amounts of feed in infrequent feeding bouts will tend to lower digestive efficiency and nutrient utilization. In a review of the effects of feeding frequency (in indoor feeding systems) on the growth and efficiency of food utilization of ruminants, Gibson (1981) found that increasing the frequency of feeding of cattle improved utilization and liveweight gain by 16%.

## FACTORS INFLUENCING HERBAGE INTAKE ON A DAILY BASIS

On a day to day basis, the main factors restricting herbage intake, are herbage availability (see earlier definition) and herbage quality.

*Grazing behaviour*

Spedding, Large and Kydd, (1966) considered the daily intake of herbage by the grazing cow to be the product of three variables;

a) the total time spent grazing per day,
b) the biting rate per minute, and
c) DM intake per bite.

Total time devoted to grazing varies between 420-700 min/day (Arnold, 1981; Fitzsimons, 1995). Biting rate generally varies in the range 45-65 bites/min (Chacon and Stobbs, 1976; Phillips, 1993; Fitzsimons, 1995), but is influenced by intake per bite, tending to increase when intake per bite is depressed. Other studies have suggested that cows are constrained to a maximum of 4000 bites per day, which effectively restricts the ability of a cow to compensate for large reductions in intake per bite (Hodgson, 1981). Consequently, it is evident that intake per bite (bite size) is the principal determinant of daily intake, and as such is sensitive to changes in sward composition and character. Stobbs (1974) measured bite size using oesophageal-fistulated animals grazing tropical pastures of varying canopy structure, and concluded that sward bulk density and the relationship between stem content and leaf:height ratio was the major factor determining bite size. Maximum biting rates have been recorded at minimum forage availability (Chacon and Stobbs, 1976), but as herbage availability increases, biting rate declines as individual bite size increases (Scarnecchai, Nastis and Malechek, 1985). Chacon and Stobbs (1976) considered the presence (i.e. acceptability) and density (i.e. accessibility) of leaf in the grazed horizon as the major factors determining daily herbage intake.

Fitzsimons (1995) calculated the bite size of British Friesian dairy cows from fistula samples collected over three or four day residency periods. Bite size varied from 0.42 to 0.81 g DM per bite, and tended to decline over time as the herbage on offer was depleted. Earlier work by Stobbs (1974) reported that a range between 0.31 and 0.71 g DM per bite is more typical of the smaller breeds such as the Jersey. Whether bite size variation is a breed characteristic, or a reflection of animal size is not known.

In more recent trials at Hillsborough, intake rates of dairy cows allowed access to swards varying in sward height and bulk density were assessed over short (1 hour) grazing periods. Data presented in Figure 8.3 illustrated the overriding importance of intake/bite in achieving high intake rates. Intake/bite increased linearly with increasing sward height between 80 and 180 mm, varying from 0.39 to 1.19 g DM/bite (Cushnahan, McGilloway, Laidlaw and Mayne, *unpublished*). Irrespective of sward height, intake per bite was greatest on the most dense sward

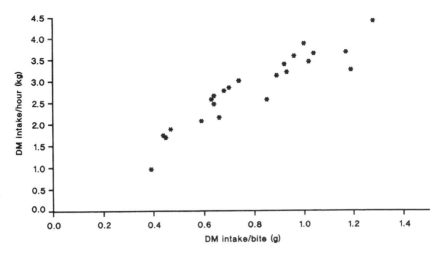

**Figure 8.3** Relationship between intake rate (kg DM/hr) and intake/bite (g DM)

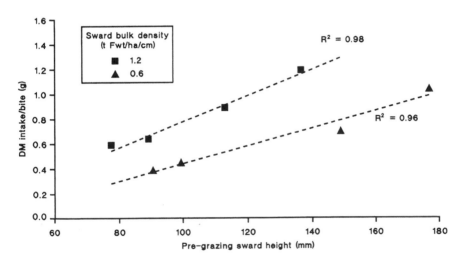

**Figure 8.4** Relationship between pre-grazing sward height (mm), DM intake/bite (g) and sward bulk density (t Fwt/ha/cm)

(Figure 8.4).The relationship between sward height, bulk density and intake (kg DM/hr) is illustrated in Figure 8.5. Intake rates in excess of 3 kg DM/hr were achieved either with a dense sward at a sward height of 120 mm, or with an open sward at approximately 155 mm. These data suggest that cows of medium genetic merit have the potential to consume upwards of 3.5 kg DM/hr. Whilst this represents the upper potential to intake rate, it suggests that cows can consume almost 0.25 of their daily requirement (15 kg DM/cow/day) within one hour when given access to dense swards of readily prehendable green leaf. This highlights

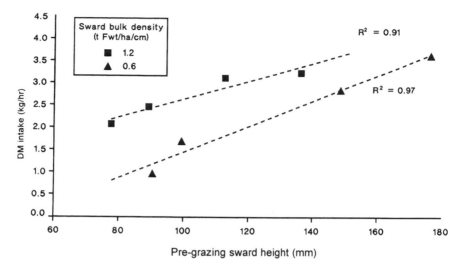

**Figure 8.5** Relationship between pre-grazing sward height (mm), DM intake (kg/hr) and sward bulk density (t Fwt/ha/cm)

the marked superiority of rotational grazing systems relative to continuous grazing systems which are generally managed at lower sward heights. It is also evident from these data that intake per bite is markedly reduced at sward heights of 80 mm relative to 180 mm (irrespective of bulk density) (Figure 8.4). From observations in the field it appears that if the herbage on offer is too tall, then the animal is presented with difficulties in manipulating the bite into the mouth, thus biting rate is reduced.

With large herbivores such as cows, plant structure is a major determinant of selection and therefore acceptability. Cows graze green leaf preferentially and even when herbage allowance is low, the proportion of green leaf in the diet remains high. For example, Fitzsimons (1995) has shown that cows spending either three or four days per residency period, still managed to maintain a leaf fraction in the diet of approximately 0.85 on the final day of residence, although this was accompanied by a small but steady increase in the dead and stem content of the diet. Other data indicate that where the physical accessibility of the leaf is reduced, energy expenditure of the grazing animal is increased (O'Reagain and Mentis, 1989). Leaf accessibility also directly affects bite size and therefore intake rates (Stobbs, 1973).

Whilst individual bites are the smallest unit of feeding behaviour, a more frequently studied unit is the meal or grazing bout (Baile and Della-Fera, 1981), which is defined as the period from when grazing commences until rest. There are generally two grazing bouts between AM and PM milking, and one grazing bout between PM milking and one to two hours after dusk. During the night, and up until AM milking there are between one and two grazing bouts.

Large animals will experience greater restrictions in bite depth and bite volume on short swards than small animals, and there is a clear effect of body size on the ability to obtain adequate energy intake from short herbage (Illius and Gordon, 1987). Cows grazing short swards are unable to eat sufficient quantities of DM, even if the area of pasture offered is large. Allden and Whittaker (1970) associated the inability of animals to graze swards of low herbage mass with reduced height, and not with any inherent characteristics of the sward such as lack of leaf material.

Combellas and Hodgson (1979), Hodgson and Jamieson (1981) and Brereton and Carton (1988) have all shown that cattle grazing swards of *L. perenne* in which the leaf material is exhausted, will reduce grazing time or even cease grazing altogether. However Wade (1991) made the point that this is just as likely to be the result of prehension difficulties, as opposed to a lack of herbage mass. However, the hypothesis that rate of intake is determined by sward height, is contradicted to some extent by Le Du, Combellas, Hodgson and Baker (1979), who concluded that intake restriction occurs when 50% of the herbage mass has been consumed. This would suggest that the residual sward height in rotational grazing systems should vary in direct proportion to the herbage mass and sward height on offer.

### Sward Structure

Sward structure describes the proportion of leaf, stem and dead material in the sward, and the relative vertical distribution of these in the sward profile. Four components can be identified. There is a horizontal separation into an upper potentially grazed horizon and a lower ungrazed horizon, and a vertical separation of the grazed and rejected area of the pasture. Swards are composed of a collection of tillers, which in turn comprise a vertical axis bearing leaves. In the flowering tiller the vertical axis is formed by stem, and in the vegetative tiller by sheath bundles. Leaf, stem and/or pseudostem length determine the potential height of the canopy, whilst leaf angle determines orientation of the photosynthetic material. Swards therefore comprise an upper leaf canopy of highly digestible material, presented over a lower layer of relatively low digestibility. Consequently, forcing cows to graze to low residual sward heights reduces the overall nutritive value of the herbage consumed.

On dense swards, Jamieson (1975 – cited by Hodgson, 1982) found an apparently linear relationship between herbage intake and sward height with grazing cattle. This was considered by Hodgson (1982) to be due to tall swards being more prehendable, but he conceded that the presence of leaf sheath in the grazed horizon may have contributed to, or caused, the decline in herbage intake that was found with declining sward height.

*Herbage dry matter content*

The DM content of the herbage on offer can have a significant effect on herbage intake rates at pasture. In a pasture diet, the nutrient content will be markedly diluted by the presence of water, which can vary from 85% in early spring to around 75% in mid-summer. This water is predominantly intracellular in nature, and thus makes a large contribution to the bulk of the diet. Evidence that voluntary intake of DM is limited by the DM content of fresh forage comes from several studies over a range of DM contents (Johns, 1954; Duckworth and Shirlaw, 1958; Arnold, 1962; Halley and Dougall, 1962; Lloyd-Davies, 1962; Moral, 1982). For dairy cows the threshold at which forage water content limits voluntary intake is thought to be between 150 and 180 g/kg DM, with an estimated depression of 0.34 kg DM intake/10 g/kg decline in DM below this level (Vérité and Journet, 1970). It is postulated that the mechanism is due to a bulk effect on rumen fill, but this is only poorly understood.

John and Ulyatt (1987) reported results of a series of studies which examined the influence of plant composition and the role of rumen fill with fresh grasses in sheep. They found a positive relationship between voluntary intake of fresh grass DM and forage DM content over a wide range (120-250 g/kg) of DM concentration, but there was no correlation between intake of fresh weight and forage DM concentration (Figure 8.6). Limitations to fresh food intake due to bulk were not attributed to negative feedback from distension of the rumen.

A system of feed intake based on the grinding and pelleting of fresh herbage is not feasible with regard to costs, but it may be possible to cut herbage at a sward height of 60 to 70 mm (thereby avoiding the bulk of stem and pseudostem material) and allow to wilt prior to grazing. Selection for herbage varieties with higher DM concentrations might also prove fruitful.

*Leaf tensile strength*

Leaf tensile strength may also be an important selection criteria influencing herbage intake in the grazing animal. Tough leaves require greater expenditure of energy to harvest (O'Reagain and Mantis, 1989). Furthermore passage of dietary residues out of the rumen requires comminution to a critical particle size (approximately 2 mm). Consequently any plant attribute that aids particle size reduction will have an impact on intake and nutritive value. Grasses are known to vary with respect to leaf strength (Evans, 1964), and it has been demonstrated that low leaf tensile and shear strength results in higher intake rates in sheep (Inoue, Brookes, John, Barry and Hunt, 1993). Both tensile and sheer strength are known to be highly

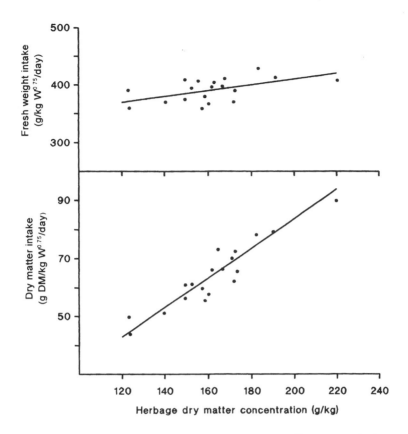

**Figure 8.6** Relationship between voluntary consumption (g/kg live wt[0.75]/day) on wet feed and dry matter basis and feed dry matter content (g/kg) for perennial ryegrass.
Source: John and Ulyatt (1987)

heritable and so offer the possibility of breeding for superior intake characteristics based on these traits (Henry, MacMillan, Roberts and Simpson, 1993). The primary tissue influencing leaf strength is the sclerenchyma bundle (John, Inoue, Brookes, Hunt and Easton, 1989), and therefore reduction of sclerenchyma tissue could reduce leaf tensile/shear strength and so increase feeding value.

Ulyatt and Waghorn (1993) also make the case for breeding for reduced protein solubility, e.g. by including condensed tannins in the plant, which will reduce the degradation of protein in the rumen and thereby improve nutritive value. Selection can also be made for cultivars with a slower rate of nutrient release during chewing (Kudo, Cheng, Hanna, Howarth, Goplen and Costerton, 1985).

**Supplementation at pasture**

The provision of supplementary feeds to dairy cows at pasture is normally undertaken either to improve animal performance over and above that which can be produced from pasture alone, or to maintain performance during periods of grass deficit. Improvement in milk yields of dairy cows at pasture (either as a result of increasing genetic merit or changes in production system, e.g. summer calving dairy herds) has meant that these considerations are becoming increasingly important.

The major challenges in supplementary feeding at pasture relate to difficulties in estimating dry matter intake and herbage composition on a daily basis. This information is an essential prerequisite in aiding decisions such as level and composition of concentrate feeds, timing of feeding and the role of supplementary forages. Furthermore, the effect of provision of supplementary feeds on herbage intake, i.e. substitution rate, is critical in determining whether production responses to supplementation are economic. However unlike the indoor feeding situation, there is relatively little information on the relationship between milk output and level of supplementary feeding at pasture.

CONCENTRATE SUPPLEMENTS

*Level of feeding*

Mayne (1991) summarized the effects of concentrate feeding on herbage substitution rates and milk production responses over a range of herbage allowances as shown in Table 8.5. These data indicate marked increases in substitution rate, and reductions in milk production responses with increases in herbage allowance.

**Table 8.5** EFFECT OF SUPPLEMENTATION AT DIFFERENT HERBAGE ALLOWANCES ON SUBSTITUTION RATE AND RESPONSE.

| Herbage allowance (kg OM[+]/day) | Substitution rate (kg herbage OM/ kg concentrate OM) | Increase in ME intake (MJ/day/kg concentrate OM) | Predicted milk response (kg/day/kg concentrate OM) |
|---|---|---|---|
| 15 | 0.11 | 12 | 1.0 |
| 20 | 0.30 | 9 | 0.8 |
| 25 | 0.50 | 7 | 0.6 |
| 30 | 0.69 | 4 | 0.4 |

[+] Organic matter

Source : Mayne (1991).

The large substitution effects observed at generous herbage allowances can be attributed to effects on grazing behaviour, with reductions in time spent grazing of 15-22 min/kg DM concentrate offered per day (Sarker and Holmes, 1974; Cowan, Byford and Stobbs, 1975). However it is interesting to speculate whether such large reductions in grazing time would be obtained at similar herbage allowances with high genetic merit cows, given the increased nutrient demand, and consequently intake drive, associated with higher levels of milk production.

Holden, Muller and Fales (1994) estimated pasture and total DM intake for high genetic merit cows grazing Orchardgrass, Kentucky-bluegrass and Smooth-broomegrass pastures in the US. Cows in mid-lactation at turnout were supplemented at the rate of 1 kg concentrate DM/5 kg milk, with a maximum of 10 kg and minimum of 4 kg of concentrate DM/day. Over the period 30 April-1 July, mean concentrate, pasture and total DM intakes averaged 7.3, 14.5 and 21.8 kg/day respectively with a mean milk yield of 33 kg milk/cow/day.

In more recent studies at Hillsborough (Ferris, Gordon, Patterson and Mayne *unpublished*) high genetic merit cows ($PTA_{90}$ fat+protein yield = 53.2 kg; sd 9.6) (on average 90 days after calving at turnout) receiving 5.5 kg concentrate fresh-weight/day produced 31.5 kg milk/d during the first 112 days at pasture. In contrast, a group of medium genetic merit cows ($PTA_{90}$ fat+protein yield = 10.9 kg; sd 10.7) receiving a similar level of concentrate, produced 26.8 kg milk/day over the same time period at pasture. These data clearly illustrate the potential to achieve higher levels of performance with high merit cows at pasture, feeding only moderate levels of concentrate supplementation, relative to that attainable with indoor housing and grass-silage based diets at similar concentrate inputs.

Again, attention is drawn to the fact that much of the research on the effect of concentrate supplementation at pasture on animal performance has been undertaken with animals of moderate milk yield potential (generally in the range 15-25 kg milk/day). Consequently, caution must be exercised in extrapolating these results to the pasture-based environment with animals of higher milk production potential.

### Concentrate allocation systems

Whilst there is no published information on the effect of concentrate allocation systems to dairy cows at pasture, extrapolation from indoor feeding studies suggests that feed-to-yield systems rather than flat rate feeding systems may be justified with concentrate feeding at pasture. This reflects the fact that in the grazing situation, sward and/or animal behaviour constraints may limit herbage intake with higher producing animals. Results of indoor feeding studies (Broster and Thomas, 1981) indicate that, where forage intake is restricted, cows of higher production potential produced a greater milk yield response to incremental increases in concentrate feeding.

Recent studies in the US (Buckmaster, Muller, Hongerholt and Gardner, 1995) have examined the effect of frequency of feeding of complementary concentrates, offered in addition to grazed pasture, on performance of high yielding (mean 38.4 kg milk/day) dairy cows.  Concentrates were offered either at pasture over four meals/day (via a mobile concentrate feeder), or twice daily at milking.  Preliminary results indicate that increased frequency of concentrate feeding had no effect on milk yield, although increases in milk fat concentration and milk fat yields were obtained at high concentrate feed levels (mean concentrate intake 12.2 kg/cow/d). However, timing of concentrate feeding may have important implications on herbage substitution rate.  For example at relatively low concentrate feed levels it may be beneficial to offer a greater proportion of concentrates after peak grazing periods (i.e. at PM rather than AM milking).

### Concentrate type

A wide range of concentrate supplements are fed to dairy cows at pasture.  These vary from supplements that are high in sugar (molasses), to starch (cereal grains) and digestible fibre (beet pulp, citrus pulp, brewers' grains, etc.).

### Energy source

The effects of concentrate energy source on substitution rate and animal performance of dairy cows at grass are inconsistent.  For example Meijs (1986) observed higher substitution rates and lower animal performance responses with high starch (350 g starch+sugars/kg DM) relative to high fiber (100 g starch+sugars/ kg DM) concentrates.  However, Van Vuuren, Van der Koelen and Vroons-de Bruin (1993) observed no effect of concentrate energy source on herbage intake, although organic matter digestion in the fore stomach was reduced and duodenal amino acid flow increased with high starch (487 g starch+sugar/kg DM) relative to high fiber (94g starch+sugar/kg DM) concentrates.  Van Vuuren *et al.*, (1993) concluded that the efficiency of microbial protein synthesis was increased with high starch concentrates, even when relatively high levels of concentrate (7.1 kg DM/day) were offered in addition to herbage.

The apparent conflict between these studies may be related to changes in the chemical composition of herbage throughout the season.  More information is required on the effect of concentrate energy source on rumen fermentation characteristics, herbage substitution rate and animal performance with a range of herbage types.

*Protein*

Whilst the crude protein content of perennial ryegrass based swards is high (particularly in spring and autumn) less than 30% of the ingested grass protein reaches the duodenum (Beever, Losada, Cammell, Evans and Haines, 1986). A high proportion of the rumen degradable protein in fresh herbage is absorbed from the rumen and excreted as urea in urine, resulting in increased potential for nitrate leaching. Consequently supplementation strategies should be designed to increase utilization of rumen degradable nitrogen by improving the synchronization of energy and protein supply to the rumen microflora. Possible options include feeding supplements with high levels of starch and sugars, or alternatively varying the timing of supplementary feeding during the day.

Recent evidence (Hongerholt, 1995) indicates that, as a result of reduced N flow to the duodenum with herbage-based diets, substantial responses in pasture intake, milk and milk protein yield can be obtained with high genetic merit cows through increases in the undegradable protein content of supplementary concentrates. The potential benefits of protected proteins arise from reduced degradation in the rumen, allowing a greater amount of dietary protein to reach the small intestine and become available to the animal (Kaufmann, 1979).

## FORAGE SUPPLEMENTS

With the increase in the levels of animal performance during the grazing season, there is a developing trend towards feeding conserved forage either once or twice daily post milking (buffer feeding), or in more extreme cases feeding forage supplements to animals housed overnight (partial storage feeding) (Phillips and Leaver, 1985).

### *Adequate grass availability*

In situations where grass supply is adequate, offering grass silage as a buffer feed results in large substitution effects (up to 0.89 kg herbage DM/kg grass silage DM), and reductions in both milk and milk protein yield (Mayne, 1991). It is important to note that substitution effects with grass silage supplementation are much greater than that with concentrate supplements (0.69 kg herbage DM/kg supplement DM) in situations where adequate herbage is available. This appears to be related to a greater depression in grazing time/day with grass silage (43 min/ d/kg forage DM) than with concentrate supplements (20-23 min/d/kg forage DM).

Maize silage has also been examined as a buffer feed for grazing dairy cows. Bryant and Donnelly (1974) observed a reduction in animal performance with buffer feeding of maize silage when adequate pasture was available. However, Holden, Muller, Lykos and Cassidy (1995), reported that feeding 2.3 kg maize silage DM/day to high yielding (29.0 kg milk/day) dairy cows grazing grass pastures and receiving 8.7 kg concentrate DM/day had no effect on milk production or total DM intake.

The data presented above suggest that in situations where adequate grass is available, shortfalls in nutrient supply from herbage, relative to nutrient requirements for high producing dairy cows, should be rectified by supplementation with complementary concentrates of high nutrient concentration, rather than forage supplements. Given that high quality grazed grass has the potential to support milk yields up to 27.0 kg/day in spring, declining to 20.0 kg/day in late August, the appropriate level of concentrate supplementation for high producing dairy cows can be calculated as shown in Table 8.6. These data relate to a rotational grazing system with a daily herbage allowance of 25.0 kg DM/cow, and assume a typical substitution rate of 0.40 kg herbage DM/kg supplement DM (Mayne, 1991). They also assume an incremental increase in herbage intake of 0.125 kg DM/kg increase in milk yield, well below the 0.4-0.5 kg incremental increase in intake normally observed in animals yielding between 15.0-25.0 kg milk/day (Stakelum, 1993; Stakelum and Dillon, 1995).

**Table 8.6** SUGGESTED CONCENTRATE FEED LEVELS FOR HIGH YIELDING DAIRY COWS IN EARLY AND LATE SEASON WITH HIGH HERBAGE ALLOWANCE (25 KG DM/COW/DAY).

|  | Mid May Target yield (kg milk/day) | | | Late August Target yield (kg milk/day) | |
| --- | --- | --- | --- | --- | --- |
|  | *25.0* | *35.0* | *40.0* | *25.0* | *35.0* |
| Potential production from grass (kg milk/day)[+] | 27.0 | 29.4 | 30.9 | 20.0 | 24.5 |
| ME required from supplement (MJ/cow/day)[++] | 0.0 | 28.0 | 45.5 | 25.0 | 52.5 |
| Supplement feed level required (kg DM/cow/day)[+++] | 0.0 | 3.9 | 6.3 | 3.5 | 7.3 |

[+]     Assumes increase in herbage intake of 0.125 kg DM/kg additional milk.
[++]    Assumes ME required for milk production of 5.0 MJ/kg milk.
[+++]   Assumes substitution rate of 0.4 kg herbage DM/kg supplement DM and ME concentration of herbage and supplement of 12.0 MJ/kg DM.

*Grass shortages*

The use of conserved forage as a supplement to grazed grass during periods of grass shortage, or when herbage quality is poor (e.g. swards in late season that have been undergrazed in spring) generally results in increased total DM intake and improved animal performance (Phillips, 1988). In these situations it is important to continue to monitor grass utilization on the grazing area, e.g. through sward height measurement, and either eliminate or restrict access to the buffer feed when sward conditions improve.

## Conclusions

Management of high genetic merit dairy cows at pasture presents a major challenge to the dairy industry in the UK and Ireland. The future profitability of milk production and milk processing will be highly dependent upon maintaining low cost, grass-based production systems. Whilst there are considerable opportunities to increase the potential contribution of grazed grass in the diet of high merit cows, full exploitation of this potential will require grazing management systems designed to present herbage to the animal in a form that will maximise intake/bite and daily DM intake. Critical factors influencing herbage intake include herbage DM content (specifically intracellular water), and sward characteristics that determine bite size. Bite size is optimized with tall, dense, leafy swards, and this can best be achieved by rotational grazing systems. Even in these systems, high merit cows will require supplementation with complementary concentrates of high nutrient concentration in order to fully exploit their genetic potential for milk production. Buffer feeding of forages and partial storage feeding systems have a role in overcoming periodic shortages in grass supply, but do not have a major role in increasing animal performance above that attainable from well managed rotationally grazed swards.

## References

Agricultural and Food Research Council (1993) *Energy and Protein Requirements of Ruminants*. An advisory manual prepared by the AFRC Technical Committee on Responses to Nutrients. CAB International, Wallingford.

Allden, W.G. and Whittaker, A.M. (1970) The determinants of herbage intake by grazing sheep : the interrelationship of factors influencing herbage intake and availability. *Australian Journal of Agricultural Research* **21** : 755-766.

Arnold, G.W. (1962) Effects of pasture maturity on the diet of sheep. *Australian Journal of Agricultural Research* **13** : 701-706.

Arnold, G.W. (1981) Grazing behaviour. In: *World Animal Science 1. Grazing Animals* pp.79-104. Ed. F.H.W. Morley. Elsevier Scientific Publishing, New York

Baile, C.A. and Della-Fera, M.A. (1981) Nature of hunger and satiety control systems in ruminants. *Journal of Dairy Science* **64** : 1140-1153.

Ball, R., Molloy, L.F. and Ross, D.J. (1978) Influence of fertilizer nitrogen on herbage dry matter and nitrogen yields, and botanical composition of a grazed grass-clover pasture. *New Zealand Journal of Agricultural Research* **21** : 47-55.

Bauman, D.E., Eppard, P.J., DeGeeter, M.J. and Lanza, G.M. (1985) Responses of high producing dairy cows to long-term treatment with pituitary somatotropin and recombinant somatotropin. *Journal of Dairy Science* **68** : 1352-1362.

Beever, D.E., Losada, H.R., Cammell, S.B., Evans, R.T. and Haines, M.J. (1986) Effects of forage species and season on nutrient digestion and supply in grazing cattle. *British Journal of Nutrition* **56** : 209-225.

Bertrand, J.A., Berger, P.J., Freeman, A.E. and Kelley, D.H. (1985) Profitability in daughters of high versus average Holstein sires selected for milk yield of daughters. *Journal of Dairy Science* **68** : 2287-2294.

Brereton, A.J. (1987) Efficient utilization of pasture requires flexible management. In: *Agricultural Pasture Improvement* Ed. L. Miro-Granada Gelabert. Commission of the European Community, Madrid, Spain. pp.147-154.

Brereton, A.J. (1995) Regional and year-to-year variation in production. In: *Irish Grasslands-their Biology and Management* pp.12-22. Eds. D.W. Jeffery, M.B. Jones, and J.H. McAdam. Royal Irish Academy, Dublin.

Brereton, A.J. and Carton, O.T. (1988) Sward height, structure and herbage use. In: *Research Meeting No. 1*, British Grassland Society, Edinburgh

Brereton, A.J., Danielov, S.A. and Scott, D. (1995) Agrometeorology of grasslands for the middle latitudes. *World Meteoroology Organisation Technical Note*, Geneva

Broster, W.H. and Thomas, C. (1981) The influence of level and pattern of concentrate input on milk output. In: *Recent Advances in Animal Nutrition 1981* pp. 49-69. Ed. W. Haresign. Butterworths, London.

Bryant, A.M. and Donnelly, P.E. (1974) Yield and composition of milk from cows fed pasture herbage supplemented with maize and pasture silages. *New Zealand Journal of Agricultural Research* **27** : 491-493.

Buckmaster, D., Muller, L., Hongerholt, D. and Gardner, M. (1995) Mobile computer controlled concentrate feeder for lactating cows. *Summaries of Pasture Research and Extension Activities* Penn State University, University Park PA.

Burnside, E.B., Graham, N., Rapitta, A.E., McBride, B.W. and Gibson, J.P. (1991) The Canadian approach to breeding efficient dairy cows. In: *Proceedings British Cattle Breeders Club Winter Conference, Cambridge* pp.25-28.

Butler-Hogg, B.W., Wood, J.D. and Bines, J.A. (1985) Fat partitioning in British Friesian cows; the influence of physiological state on dissected body composition. *Journal of Agricultural Science*, Cambridge **104** : 519-528.

Chacon, E. And Stobbs, T.H. (1976) Influence of progressive defoliation of a grass sward on the eating behaviour of cattle. *Australian Journal of Agricultural Research* **27** : pp.709-718.

Combellas, J. and Hodgson, J. (1979) Herbage intake and milk production by grazing dairy cows 1. The effect of variation in herbage mass and daily herbage allowance in a short-term trial. *Grass and Forage Science* **34** : 209-214.

Connaughton, M.J. (1973) The grass-growing season in Ireland. *Agrometeorological Memorandum No. 5.* Dublin Meteorological Service.

Cowan, R.T., Byford, I.J.R. and Stobbs, T.H. (1975) Effects of stocking rate and energy supplementation on milk production from tropical grass-legume pasture. *Australian Journal of Experimental and Animal Husbandry* **15** : 740-746.

Duckworth, J.E. and Shirlaw, D.W. (1958) A study of factors affecting feed intake and the eating behaviour of cattle. *Animal Behaviour* **6** : 147-154.

Evans, P.S. (1964) A study of leaf strength in four ryegrass varieties. *New Zealand Journal of Agricultural Research* **7** : 508-513.

Evans, B. (1981) Production from swards grazed by dairy cows. *Grass and Forage Science* **36** : 132-134.

Fitzsimons, E. (1995) Influence of sward structure on the grazing behaviour of dairy cows. *M.Sc. Thesis*, University College Cork, Ireland.

Flores, E.R., Laca, E.A., Griggs, T.C. and Demment, M.W. (1993) Sward height and vertical morphological differentiation determine cattle bite dimensions. *Agronomy Journal* **85** : 527-532.

Forbes, J.M. (1995) Voluntary intake - A limiting factor to production in high-yielding dairy cows? In: *Breeding and Feeding the High Genetic Merit Cow* British Society of Animal Science Occasional Publication No. 19. pp.13-19.

Funk, D.A. (1993) Optimal genetic improvement for the high producing herd. *Journal of Dairy Science* **76** : 3278-3286.

Gibb, M.J., Ivings, W.E., Dhanoa, M.S. and Sutton, J.D. (1992) Changes in body components of autumn calving Holstein-Friesian cows over the first 29 weeks of lactation. *Animal Production* **55** : 339-360.

Gibson, J.P. (1981) The effects of feeding frequency on the growth and efficiency of food utilization of ruminants : An analysis of published results. *Animal Production* **32** : 275-283.

Gill, M. And Beever, D.E. (1991) Modeling as an aid to nutrition research. In: *Isotope and related techniques in animal production and health* pp.171-182. IEAE, Vienna.

Graham, N.J., Burnside, E.B., Gibson, J.P., Rapitta, A.E. and McBride, B.W. (1991) Comparison of daughters of Canadian and New Zealand Holstein sires for first lactation efficiency of production in relation to body size and condition. *Canadian Journal of Animal Science* **71** : 293-300.

Grainger, C., Holmes, C.W. and Moore, Y.F. (1985) Performance of Friesian cows with high and low breeding indexes 2. Energy and nitrogen balance experiments with lactating and pregnant, non lactating cows. *Animal Production* **40** : 389-400.

Grant, S.A., Barthram, G.T. and Torvell, L. (1981) Components of regrowth in grazed and cut *Lolium perenne* swards. *Grass and Forage Science* **36** : 155-168.

Halley, R.J. and Dougall, B.M. (1962) The feed intake and performance of dairy cows fed on cut grass. *Journal of Dairy Research* **29** : 241-248.

Hansen, L.B., Freeman, A.E. and Berger, P.J. (1983) Yield and fertility relationships in dairy cattle. *Journal of Dairy Science* **66** : 293-305.

Henry, D.A., MacMillan, R.H., Roberts, F.M. and Simpson, R.J. (1993) Assessment of the variation in shear strength of leaves of pasture grasses. *Proceedings 17th International Grassland Congress*, Palmerston North, New Zealand. pp.533-555.

Hodgson, J. (1981) Variations in the surface characteristics of the sward and the short-term rate of herbage intake by calves and lambs. *Grass and Forage Science* **36** : pp.49-57.

Hodgson, J. (1982) Influence of sward characteristics on diet selection and herbage intake by the grazing animal. In: *Nutritional Limits to Animal Production from Pasture* pp.153-166, Commonwealth Agricultural Bureaux, St Lucia, Australia.

Hodgson, J. and Jamieson, W.S. (1981) Variations in herbage mass and digestibility, and the grazing behaviour of adult cattle and weaned calves. *Grass and Forage Science* **36** : 39-48.

Holden, L.A., Muller, L.D. and Fales, S.L. (1994) Estimation of intake in high producing Holstein cows grazing grass pasture *Journal of Dairy Science* **77** : 2332-2340.

Holden, L.A., Muller, L.D., Lykos, T. And Cassidy, T.W. (1995) Effect of corn silage supplementation on intake and milk production in cows grazing grass pasture. *Journal of Dairy Science* **78** : 154-160.

Holmes, C.W. (1988) Genetic merit and efficiency of milk production by the dairy cow. In: *Nutrition and Lactation in the Dairy Cow* pp.195-215. Ed. P.C. Garnsworthy, Butterworths, London.

Holmes, C.W. (1995) Genotype x environment interactions in dairy cattle: a New Zealand perspective. In: *Breeding and Feeding the High Genetic Merit Cow* British Society of Animal Science Occasional Publication No. 19. pp.51-58, Edinburgh

Holmes, W. (1968) The use of nitrogen in the management of pasture for cattle. *Herbage Abstracts* **38** : 265-277.

Hongerholt, D.D. (1995) Grain supplementation strategies for dairy cows grazing grass pastures and their effects on milk production and microbial fermentation. *Ph.D. Thesis, Pennsylvania State University.*

Illius, A.W. and Gordon, I.J. (1987) The allometry of food intake in grazing ruminants. *Journal of Animal Ecology* **56** : 989-999.

Illius, A.W., Lowman, B.G.and Hunter, E.A. (1986) The use of buffer grazing to maintain sward quality and increase late season cattle performance. In: *Grazing* pp.119-128. Ed. J. Frame. British Grassland Society Occasional Symposium No. 19, BGS, Hurley.

Inoue, T., Brookes, I.M., John, A., Barry, T.N. and Hunt, W.F. (1993) Effect of physical resistance in perennial ryegrass leaves on feeding value for sheep. *Proceedings 17th International Grassland Congress*, Palmerston North, New Zealand. pp.570-572.

Jamieson, W.S. (1975) Studies on the herbage intake and grazing behaviour of cattle and sheep. *Ph.D. Thesis*, University of Reading.

John, A., Inoue, T., Brookes, I.M., Hunt, W.F. and Easton, H.S. (1989) Effects of selection for shear strength on structure and rumen digestion of perennial ryegrass. *Proceedings of the New Zealand Society of Animal Production* **49** : 225-228.

John, A. and Ulyatt, M.J. (1987) Importance of dry matter content to voluntary intake of fresh grass forages *Proceedings of the New Zealand Society of Animal Production* **47** : 13-16.

Johns, A.T. (1954) Bloat in cattle on red clover. *New Zealand Journal of Science and Technology* **36** : 289-320.

Journet, M. and Demarquilly, C. (1979) Grazing In: *Feeding strategies for the high yielding dairy cow* pp.295-321. Eds. W.H. Broster and H. Swan. Granada Publishing, London.

Kaufmann, W. (1979) In: *Feeding Strategies for the High Yielding Dairy Cow* pp.90-113. Eds. W.H. Broster and H. Swan. Granada Publishing, London

Kingsbury, L.R. (1965) Pasture quality in terms of soluble carbohydrates and volatile fatty acid production. *Proceedings of the New Zealand Society of Animal Production* **25** : 119-136.

Kudo, H., Cheng, K.J., Hanna, M.R., Howarth, R.E., Goplen, B.P. and Costerton, J.W. (1985) Ruminal digestion of alfalfa strains selected for slow and fast initial rates of digestion. *Canadian Journal of Animal Science* **65** : 157-161.

Laca, E.A., Demment, M.W., Distel, R.A. and Griggs, T.C. (1993) A conceptual model to explain variation in ingestive behaviour within a feeding patch. *Proceedings 17th International Grassland Congress*, Palmerston North, New Zealand. pp.710-712.

Laca, E.A., Ungar, E.D., Seligman, N.G. and Demment, M.W. (1992) Effects of sward height and bulk density on bite dimensions of cattle grazing homogenous swards. *Grass and Forage Science* **47** : 91-102.

Leaver, J.D. (1983) Feeding for high margins. In: *Recent Advances in Animal Nutrition - 1983* pp.199-207. Ed. W. Haresign. Butterworths, London.

Leaver, J.D. (1985) Milk production from grazed temperate grassland. *Journal of Dairy Science* **52** : 313-344.

Leaver, J.D. (1987) The potential to increase production efficiency from animal-pasture systems. *Proceedings of the New Zealand Society of Animal Production* **47** : 7-12.

Le Du, Y.L.P., Combellas, J., Hodgson, J. and Baker, R.D. (1979) Herbage intake and milk production by grazing dairy cows 2. The effects of level of winter feeding and daily herbage allowance. *Grass and Forage Science* **34** : 249-260.

Lloyd-Davies, H. (1962) Intake studies in sheep involving high fluid intake. *Proceedings of the Australian Society of Animal Production* **4** : 167-171.

Mayne, C.S. (1991) Effects of supplementation on the performance of both growing and lactating cattle at pasture. In: *Management Issues for the Grassland Farmer in the 1990s* pp.55-71. Ed. C.S. Mayne. Occasional Symposium No. 25. British Grassland Society, Hurley.

Mayne, C.S. and Gordon, F.J. (1995) Implications of genotype x nutrition interactions for efficiency of milk production systems. In: *Breeding and Feeding the High Genetic Merit Cow* British Society of Animal Science Occasional Publication No. 19. Edinburgh pp.67-77.

McMeekan, C.P. and Walsh, M.J. (1963) The inter-relationships of grazing method and stocking rate in the efficiency of pasture utilisation by dairy cattle. *Journal of Agricultural Science* **61** : 147-166.

Meijs, J.A.C. (1986) Concentrate supplementation of grazing dairy cows 2. Effect of concentrate composition on herbage intake and milk production. *Grass and Forage Science* **41** : 229-235.

Meijs, J.A.C. and Hoekstra, J.A. (1984) Concentrate supplementation of grazing dairy cows 1. Effect of concentrate intake and herbage allowance on herbage intake. *Grass and Forage Science* **61** : 147-166.

Moral, M.M.C. del (1982) Study of the nutritive value of a mixed pasture: variations in digestibility and intake. *Pastos* **12** : 119-133. (Summarised in *Nutrition Abstracts and Reviews (series B)* **56** (4)).

Morrison, J. (1977) The growth of *Lolium perenne* and response to fertilizer-N in relation to season and management. *Proceedings XIII International Grassland Congress,* Leipzig. pp.943-946.

Morrison, J., Jackson, M.V. and Williams, T.E. (1974) Variation in the response of grass to fertilizer N in relation to environment. *Proceedings of the Fertilizer Society* **142** : 28-38.

Morrison, J., Jackson, M.V. and Sparrow, P.E. (1980) The response of perennial ryegrass to fertilizer nitrogen in relation to climate and soil : report of the joint ADAS/GRI grassland manuring trial GM20 *Technical Report No. 27* Grassland Research Institute, Hurley.

Oltenacu, P.A., Frick, A. and Lindhe, B. (1991) Relationship of fertility to milk yield in Swedish cattle. *Journal of Dairy Science* **74** : 264-268.

O'Reagain, P.J. and Mentis, M.T. (1989) The effect of plant structure on the acceptability of different grass species to cattle. *Journal of the Grassland Society of South Africa* **6** : 163-170.

Patterson, D.C., Gordon, F.J., Mayne, C.S. Porter, M.G. and Unsworth, E.F. (1995) The effects of genetic merit on nutrient utilization in lactating dairy cows. In: *Breeding and Feeding the High Genetic Merit Cow* British Society of Animal Science Occasional Publication No. 19. pp.97-98.

Peterson, R. (1988) Comparison of Canadian and New Zealand sires in New Zealand for production, weight and conformation traits (Unpublished Report). 12pp.

Phillips, C.J.C. (1988) The use of conserved forage as a supplement for grazing dairy cows. *Grass and Forage Science* **43** : 215-230.

Phillips, C.J.C. (1993) *Cattle Behaviour* Farming Press Books, Ipswich.

Phillips, C.J.C. and Leaver, J.D. (1985) Supplementary feeding of forage to grazing dairy cows. 2. Offering grass silage in early and late season. *Grass and Forage Science* **40** : 193-199.

Reid, D. (1978) The effect of frequency of defoliation on the yield response of a perennial ryegrass sward to a wide range of nitrogen application rates. *Journal of Agricultural Science, Cambridge* **90** : 447-457.

Ryan, D.P., Snijders, S., McGilloway, D.A. and O'Farrell, K.J. (1996) Effects of grazing pressure on dry matter intake, ovarian follicular populations, and reproductive performance in lactating cows. *Journal of Animal Science* (submitted).

Sarker, H.B. and Holmes, W. (1974) The influence of supplementary feeding on the herbage intake and grazing behaviour of dry cows. *Journal of the British Grassland Society* **29** : 141-143.

Scarnecchia, D.L., Nastis, A.S. and Malechek, J.C. (1985) Effects of forage availability on grazing behaviour of heifers. *Journal of Range Management* **38** : 177-180.

Sejrsen, K. and Neimann-Sorensen, A. (1994) Sources of genetic variation in efficiency of cattle : biological basis. *Proceedings 45th Meeting of European Association of Animal Production,* Edinburgh pp 140.

Shanks, R.D., Freeman, P.J. Berger, P.J. and Kelley, D.H. (1978) Effects of selection for milk production on reproductive and general health of the dairy cow. *Journal of Dairy Science* **61** : 1765-1772.

Spedding, C.R.W., Large, R.V. and Kydd, D.D. (1966) The evaluation of herbage species by grazing animals. *10th International Grassland Congress,* Helsinki pp.479-483.

Stakelum, G. (1993) Achieving high performance from dairy cows on grazed pastures. *Irish Grassland and Animal Production Association Journal* **27** : 9-18.

Stakelum, G. and Dillon, P. (1995) Supplementary feeding of grazing dairy cows. *Technical Bulletin, Issue No. 2,* R & H Hall , Dublin

Stewart, T.A. (1985) Utilising white clover in grass based animal production systems. In: *Forage Legumes* pp.93-103. Ed. D.J. Thomson. Occasional Symposium No. 16, British Grassland Society, Hurley.

Stobbs, T.H. (1973) The effect of plant structure on the intake of tropical pastures 2. Differences in sward structure, nutritive value and bite size of animals grazing *Setaria anceps* and *Chloris gayana* at various stages of growth. *Australian Journal of Agricultural Research* **24** : 821-829.

Stobbs, T.H. (1974) Rate of biting by Jersey cows as influenced by the yield and maturity of pasture swards. *Tropical Grasslands* **8** : 81-86.

Struth, J.W., Brown, J.R., Olson, P.D., Araujo, M.R. and Alijoe, H.D. (1987) Effects of stocking rate on critical plant animal interactions in a rotationally grazed *Schizachyrium-Paspalum savannah.* In: *Grazing-lands Research at the Plant-animal Interface* pp.115-139. Eds. F.P. Horn, J. Hodgson, J.J. Mott and R.W. Brougham. Winrock International, Morriton, Arkansas, USA

Thomson, D.J., Beever, D.E., Haines, M.J., Cammell, S.B., Evans, R.T., Dhanoa, M.S. and Austin, A.R. (1985) Yield and composition of milk from Friesian cows grazing either perennial ryegrass or white clover in early lactation. *Journal of Dairy Research* **52** : 17-31.

Thran, P.S. and Broekhuizen, S. (1965) *Agro-climatic Atlas of Europe* Pudoc., Wageningen. Elsevier, Amsterdam.

UK Statistics for Genetic Evaluation (1995) Animal Data Centre Ltd, Rickmansworth

Ulyatt, M.J. and Waghorn, G.C. (1993) Limitations to high levels of dairy production from New Zealand pastures. In: *Improving the Quality and Intake of Pasture Based Diets for Lactating Dairy Cows* pp.11-32. Eds. N.J. Edwards and W.J. Parker. Department of Agricultural and Horticultural

Systems Management, Occasional Publication No. 1, Massey University, Palmerston North, New Zealand.

Van Vuuren, A.M., Van der Koelen, C.J. and Vroons-de Bruim, J. (1993). Ryegrass versus corn starch or beet pulp fibre diet effects on digestion and intestinal amino acids in dairy cows. *Journal of Dairy Science* **76** : 2692-2700.

Veerkamp, R.F., Emmans, G.C., Cromie, A.R. and Simm, G. (1995) Variance components for residual feed intake in dairy cows. *Livestock Production Science* **41** : 111-120.

Veerkamp, R.F., Simm, G. and Oldham, J.D. (1994) Effects of interaction between genotype and feeding system on milk production, feed intake, efficiency and body tissue mobilization in dairy cows. *Livestock Production Science* **39** : 229-241.

Vérité, R. And Journet, M. (1970) Influence de la teneur en eau et la deshydration de l'herbe sur la valeur alimentaire pour les vaches laitiers. *Annales de Zootechnie* **19** : 255-268.

Wade, M.H. (1991) Factors affecting the availability of vegetative *Lolium perenne* L. to grazing dairy cows with special reference to sward characteristics, stocking rate and grazing method. Thése de Docteur Ingénieur, Université de Rennes.

Wilkinson, J.M., Price, W.R., Russell, S.R. and Jones, P. (1994) Diurnal variation in dry matter and sugar content of ryegrass. *British Grassland Society, 4th Research Conference*, University of Reading. pp.61-62.

**9**

## SUPPLEMENTATION OF MAIZE SILAGE AND WHOLECROP CEREALS

J.D. Leaver
*Wye College, University of London, Ashford, Kent TN25 5AH, UK*

## Introduction

The proportion of forages fed to dairy cows in the UK as maize silage or wholecrop cereals is increasing annually at a substantial rate. This increase is mainly at the expense of grass silage, and several reasons appear to explain this change in nutritional management. In the case of forage maize, plant breeders have been successful in producing early maturing varieties which have a high probability of achieving 300g DM/kg at harvest. As a result, maize can now be successfully grown in a much wider area of the country. Wholecrop cereals are often higher yielding than maize or grass, especially autumn-sown varieties, and can be grown over a wide geographical area. They also have the flexibility of being harvested either as wholecrop or combine harvested for grain. Nevertheless, the two major reasons for the expansion of these forage crops are probably their lower cost of production, and their higher feeding value than grass silage (Phipps, 1994).

These developments in maize silage and wholecrop cereal use have coincided with a rapid rise in the genetic merit of the UK dairy herd, resulting from an increased use of proven bull semen from North American origins. This represents a further factor encouraging farmers to switch from grass silage to forages with high nutritive value and feed intake characteristics.

Supplementation of forage maize and wholecrop cereals with concentrates can have a number of objectives. Firstly, the nutritional deficiencies of the forages must be overcome. It is well established that their protein and mineral/vitamin contents are low compared with grass silages, and these deficiencies must be addressed in supplementation strategies. Secondly, supplementation aims to increase total energy and protein supply to the lactating animal in order to increase milk production. The extent of the milk production response primarily depends on the substitution rate of concentrates for forage. However, the energy source, protein source and protein content of the supplement is also likely to be influential.

In general, maize silage and wholecrop cereals are not fed as the sole forage to dairy cattle. They are normally offered in conjunction with grass silage. This can make the quantification of responses to supplementary feeding more complex. There are many challenges therefore for farmers, extension workers and researchers in developing successful nutritional management of high genetic merit cows. Understanding the interrelationships between supplements and forages containing maize silage and wholecrop cereals will be a crucial component of success.

## Feeding value of maize silage and wholecrop cereals

NUTRITIVE VALUE

Forage maize and wholecrop cereals have different nutritional characteristics from grass silage, mainly due to the different botanical components of the crops. Grass silage consists predominantly of leaf and stem, whereas grain is an important component of both maize silage and wholecrop cereal forage. Ensiled maize normally has over 500g/kg DM of cob in the total plant if harvested at 300 g DM/ kg or more (Table 9.1). Wholecrop wheat has a lower proportion of ear in the total plant, the proportion varying with stage of growth at harvest. The level is in the range 200 to 350 g/kg DM when cut at the ensiling stage (350 to 450 g DM/kg), and 350 to 500 g/kg DM when harvested for urea treatment (450 to 550 g DM/ kg). Wholecrop barley is normally harvested for ensiling at 350 to 450 g DM/kg, and the proportion of ear at that time is 350 to 450 g/kg DM (Hargreaves, 1993).

In addition to these differences between maize and wholecrop cereals in the proportion of grain in the total DM, there is also a difference in the digestibility of the vegetative components (Table 9.1). The neutral detergent cellulase digestibility (NCD) of leaf and stem tissue is higher for forage maize than for wholecrop cereals harvested at over 400 g DM/kg. This is probably due to the grain in maize reaching maturity before much of the leaf and stem has senesced, unlike small grain cereals where the leaf and stem tissue is brown at the time of harvesting. Plant breeders of maize are now producing early maturing varieties which attain the 300 g DM/kg target for harvesting when the leaves are still green. The combination of a higher grain proportion and a higher digestibility of leaves and stem of maize leads to a significantly higher digestibility and ME value of the crop than wholecrop wheat and barley (Table 9.2).

**Table 9.1** COMPONENTS OF MAIZE AND WHOLECROP CEREALS

| Forage | Storage system | Crop DM (g DM/kg) | Proportion (g/kg DM) | | | NCD (g/kg DM) | |
|---|---|---|---|---|---|---|---|
| | | | Cob/ ear | Leaf+ sheath | Stem | Leaf+ sheath | Stem |
| Maize[1] | ensiled | 317 | 545 | 256 | 222 | 610 | 536 |
| Wholecrop wheat[2] | ensiled | 372 | 260 | 322 | 418 | 602 | 530 |
| Wholecrop wheat[2] | urea | 555 | 440 | 220 | 340 | 522 | 411 |

[1] Mean of four varieties (Leaver, 1994 unpublished)
[2] Variety Haven (Leaver and Hill, 1992 unpublished).

**Table 9.2** CHEMICAL COMPOSITION OF MAIZE SILAGE, WHOLECROP WHEAT AND WHOLECROP BARLEY, EACH HARVESTED AT TWO STAGES OF MATURITY

| | Maize[1] | | Wholecrop wheat[2] | | Wholecrop barley[3] | |
|---|---|---|---|---|---|---|
| | low DM | high DM | silage | 4% urea | low DM | high DM |
| Toluene DM (g/kg) | 260 | 356 | 372 | 555 | 362 | 464 |
| CP (g/kg DM) | 86 | 79 | 91 | 152 | 80 | 73 |
| NDF (g/kg DM) | 401 | 367 | 428 | 459 | 512 | 516 |
| ADF (g/kg DM) | 271 | 223 | 264 | 309 | 300 | 279 |
| Starch (g/kg DM) | 93 | 239 | 23 | 258 | 223 | 258 |
| NCD (g/kg DM) | 689 | 698 | 603 | 619 | 583 | 592 |
| ME (MJ/kg DM)* | 10.7 | 10.8 | 9.8 | 10.0 | 9.6 | 9.5 |
| pH | 3.7 | 3.8 | 4.0 | 6.8 | 4.2 | 4.5 |
| Ammonia N (g/kg N) | 52 | 51 | 108 | 226 | 126 | 113 |

ME = 0.157 x DOMD.  DOMD = 0.65 NCD + 235 (Givens *et al.* 1995)
[1] Leaver (1993 unpublished). [2] Leaver and Hill (1995). [3] Hargreaves (1993)

These alternative forages to grass silage are characterised by low crude protein (CP), mineral and vitamin contents. The starch content is dependent on the stage of growth of the crop at harvest. In maize, wheat and barley, this can range from nil for crops harvested at the vegetative stage of growth, to over 250 g/kg DM for mature crops of wheat and barley, and 350 g/kg DM for maize. These crops have a low buffering capacity, and consequently when ensiled reach a low pH with a stable fermentation without the need for acid or acid-stimulating silage additives, providing adequate water soluble carbohydrate is present. This applies to crops up to about 450 g DM/kg, above which water soluble carbohydrate is likely to be limiting for an appropriate fermentation to take place. The low buffering capacity also means that following opening, the silages are very susceptible to aerobic deterioration as the pH can increase rapidly when exposed to air.

The reduced fermentation which occurs with advancing maturity may lead to a reduction in the proportion of nitrogen (N) which is degraded to non-protein nitrogen (Wilkinson, 1978). For all these crops the effective degradability of nitrogen (N) is high at 0.75 to 0.85 (Alderman and Cottrill, 1993).

The ammonia N content is high in urea treated wholecrop wheat (Table 9.2). Leaver and Hill (1992) suggested that the high ammonia intake where urea-treated wholecrop wheat was the sole forage, might be the cause of an apparent inefficiency of energy utilisation of the crop. High blood urea levels (over 60 mg/100 ml blood) were found, and energy balances indicated that the estimated ME intakes were substantially higher than the ME accounted for in animal output. For this reason, maximum inclusion rates in diets of 5 kg DM/day are normally recommended where wholecrop wheat is treated with 40 g urea/kg DM. Recent research by Sutton, Abdalla, Phipps, Cammell and Humphries (1995) did not find this energy inefficiency when urea-treated wholecrop wheat was the sole forage for dairy cows. Givens, Moss and Adamson (1993) have indicated however that the ME of urea-treated wholecrop wheat is equal to 0.0143 DOMD compared with 0.0157 for grass silages, which is an indication of inefficient conversion of DE to ME.

An environmental implication with urea-treated wholecrop is that an application rate of 40 g urea/kg wholecrop DM, represents an N input equivalent to approximately 250 kg N/ha in addition to a typical N fertilizer input to the crop of 150 to 200 kg N/ha. A lower rate of application of 20 g urea/kg wholecrop DM has been investigated (Leaver and Hill, 1995), and found to give an equivalent level of dairy cow performance to a 40 g urea/kg wholecrop DM application rate. It is important that this lower rate of urea is spread evenly throughout the forage.


ROLE OF STARCH

Unlike grass silages, maize silage and wholecrop cereals contain substantial proportions of starch, which is a valuable energy source for the dairy cow. However, the species of crop and the stage of growth at harvest not only influence the quantity of starch, but also substantially affect the rate and extent of starch degradation in the rumen (De Visser, 1993). This has potential consequences for DM intake, FME supply, microbial growth, milk fat and milk protein production (Kung, Tung and Carmean, 1992).

De Visser (1993) reported that for maize silage an increase in DM from 213 to 323 g DM/kg, increased the proportion of starch, resistant to rumen degradation, from 70 to 320 g/kg, and there were differences between maize varieties in this attribute. The same author reviewed research which showed that maize starch had a comparatively low rate of degradation in the rumen and as a consequence 420 g/ kg starch by-passed the rumen, compared with 80 g/kg for wheat starch and 70 g/ kg for barley starch. Thus there are likely to be differences between maize silage, and wholecrop cereals in FME supply, and potentially in microbial protein production and milk protein content and yield.

The whole tract digestibility of starch is normally greater for maize silage than for wholecrop cereals (Table 9.3). The complete digestion of starch shown for ensiled wholecrop wheat in Table 9.3 has little nutritional significance as the amount of starch in the crop is small. If the crop is very mature the digestibility can be substantially reduced. Sutton *et al.* (1995) with urea-treated wholecrop wheat of 760 g DM/kg reported a starch digestibility of only 755 g/kg. Farmers are naturally concerned about the sight of cereal grain in cow faeces. The present evidence suggests that providing maize silage is harvested below 350 g DM/kg, or a corn cracker is used on the forage harvester if above 350 g DM/kg then high starch digestibility will be attained. For wholecrop wheat a starch digestibility of over 850 g/kg is likely, providing the DM at harvest does not exceed 550 g DM/ kg. Wholecrop barley should be ensiled at a stage of maturity not exceeding 450 g DM/kg due to the hardness of the grain. Urea treatment at a higher DM is not appropriate due to the high probability of intact grain passing undigested through the alimentary tract.

**Table 9.3** DIGESTIBILITY IN HOLSTEIN FRIESIAN HEIFERS OF STARCH AND FIBRE IN MAIZE SILAGE WHOLECROP WHEAT AND WHOLECROP BARLEY, EACH HARVESTED AT TWO STAGES OF MATURITY

|  | *Maize*[1] | | *Wholecrop wheat*[2] | | *Wholecrop barley*[3] | |
|  | *low DM* | *high DM* | *silage* | *4% urea* | *low DM* | *high DM* |
|---|---|---|---|---|---|---|
| Forage DM (g/kg) | 255 | 353 | 372 | 549 | 323 | 452 |
| Starch intake (kg DM/day) | 0.84 | 2.41 | 0.13 | 1.70 | 1.54 | 2.01 |
| Starch digestibility (g/kg) | 979 | 947 | 1000 | 888 | 880 | 860 |
| NDF intake (kg DM/day) | 3.61 | 3.71 | 2.48 | 3.03 | 3.53 | 4.02 |
| NDF digestibility (g/kg) | 685 | 669 | 637 | 643 | 605 | 551 |

[1] Leaver (1993 unpublished). [2] Leaver and Hill (1995). [3] Hargreaves (1993)

ROLE OF NDF

The proportion of NDF (cell wall) in these alternative forages is not dissimilar to grass silage of moderate to high digestibility. However, as the crops mature (Table 9.2) the NDF content does not rise substantially as for grass, and may even decline due to the diluting effect of the cob/ear development. The digestibility of the NDF (Table 9.3) normally declines with advancing stage of maturity of the crop, and tends to be higher for maize than for wheat and barley due to the leaves and stem being less mature at harvest.

Information on NDF rumen degradability is of importance because it is influential on FME supply and microbial protein production in the same way as for starch (De Visser, 1993; Overton, Cameron, Elliot, Clark and Nelson, 1995). In addition NDF is an important factor in the control of intake as discussed below.

FORAGE INTAKE

The performance of dairy cattle is strongly influenced by those characteristics of the feed which influence intake. The main reason why maize silage and to some extent wholecrop cereals are advantageous nutritionally compared with grass silage is their high intake characteristics, as is clearly shown in Table 9.4. In mixed forage diets containing 330 to 400 g/kg of maize silage or wholecrop wheat in the total forage, DM intake was increased in all cases compared with grass silage alone. The increases ranged from 3 to 23%, and at higher levels of inclusion of these forages, the advantages in total forage intake can be even greater (Phipps, 1994).

**Table 9.4** EFFECTS OF FORAGE TYPE ON FEED INTAKE IN DAIRY COWS

| (kg DM/day) | Grass silage | | Maize silage | | Ensiled wholecrop wheat | | Urea treated wholecrop wheat | |
|---|---|---|---|---|---|---|---|---|
| 1. Concentrates | 6.9 | | | | 6.9 | | 6.9 | |
| Forage | 9.4 | (100) | | | 11.1 | (118) | 11.6 | (123) |
| 2. Concentrates | 6.2 | | | | 6.2 | | 6.2 | |
| Forage | 11.7 | (100) | | | 12.1 | (103) | 12.5 | (107) |
| 3. Concentrates | 6.0 | | 6.0 | | 6.0 | | 6.0 | |
| Forage | 9.3 | (100) | 10.6 | (114) | 10.6 | (114) | 10.2 | (110) |

1. Leaver and Hill (1995) - mixed forages 40:60 ratio with grass silage.
2. Leaver and Hill (1995) - mixed forages 33:67 ratio with grass silage.
3. Phipps *et al.* (1995) - mixed forages 33:67 ratio with grass silage

There are a number of factors relating to the characteristics of the forage which influence voluntary intake. The higher intakes of maize silage and wholecrop cereals compared with grass silage probably relate both to their botanical characteristics and to their fermentation characteristics in the rumen. Hill and Leaver (1993) reported estimated rumen outflow rates of 0.049 and 0.059/h for grass silage and urea-treated wholecrop wheat respectively. It seems likely therefore that due to the grain content, and possibly due to the higher DM content, these forages have a lower retention time in the rumen, thus allowing a higher forage intake. The lower buffering capacity of these forages when ensiled also means that following consumption, their pH will rise more quickly to the neutral pH of the rumen compared with grass silage. Nevertheless, there are likely to be differences between maize silage and wheat or barley silages in their subsequent effects on rumen pH following fermentation. Maize grain is less degradable in the rumen than small grain cereals and is less likely to depress rumen pH (De Visser, 1993; Kung *et al.* 1992). The negative effect of starch fermentation in the rumen on intake may therefore be less for maize than for wheat or barley silages. Urea-treated wholecrop cereals are alkaline and therefore have no negative implications for rumen pH, and this may be a useful attribute in rations containing a high proportion of cereals.

## MIXED FORAGES AND MILK PRODUCTION

The mixing of maize silage and wholecrop cereals with grass silage has been reviewed by Phipps (1994). He concluded that the 'improvement in forage intake and the resultant increased yield of milk and milk constituents' should enable farmers to 'replace a significant proportion of the grass silage ration with an appropriate complementary forage resource'.

Recent research (Table 9.5) tends to confirm these conclusions. The higher intakes achieved when maize silage or wholecrop wheat were mixed with grass silage (Table 9.4), were in general translated into higher levels of milk yield, although no consistent benefits in milk composition were found. The extent of these responses depends to a large extent on the relative nutritive values of the grass silages and alternative forages used.

The costs of production of maize and wholecrop cereals have been found to be lower than for grass silage (Leaver, 1991; Phipps, 1994), particularly when in-silo losses are also taken into account. This factor, combined with the higher feeding value of these forages compared with grass silage, makes them increasingly attractive for use by dairy farmers as a partial replacement for grass silage. However, a knowledge of responses to supplements is an integral part of nutritional management, and for these forages much less is known than for grass silage diets.

**Table 9.5** EFFECTS OF FORAGE TYPE ON MILK PRODUCTION IN DAIRY COWS

|  | Grass silage | Maize silage | Maize silage (H) | Ensiled wholecrop wheat | Untreated wholecrop wheat |
|---|---|---|---|---|---|
| 1. Milk yield (kg/day) | 28.0 | | | 28.7 | 29.6 |
| Fat (g/kg) | 40.4 | | | 40.5 | 39.8 |
| Protein (g/kg) | 31.3 | | | 31.3 | 31.2 |
| 2. Milk yield (kg/day) | 30.0 | | | 29.1 | 29.9 |
| Fat (g/kg) | 41.9 | | | 41.0 | 41.4 |
| Protein (g/kg) | 32.5 | | | 31.9 | 32.5 |
| 3. Milk yield (kg/day) | 23.0 | 26.4 | 27.6 | 24.2 | 24.0 |
| Fat (g/kg) | 41.7 | 41.8 | 40.6 | 41.7 | 42.1 |
| Protein (g/kg) | 29.9 | 31.2 | 31.9 | 30.8 | 30.8 |

1 Leaver and Hill (1995) - mixed forages 40:60 ratio with grass silage.
2 Leaver and Hill (1995) - mixed forages 33:67 ratio with grass silage.
3 Phipps *et al.* (1995) - mixed forages 33:67 ratio with grass silage or 75:25 ratio with grass silage (Maize silage (H))

## Supplementation of maize silage and wholecrop cereals

PROTEIN LEVEL AND SOURCE IN SUPPLEMENT

The protein content of maize silage and wholecrop cereals is low (Table 9.2). It is well established that the amount and type of protein supplement influences the supply of amino acids to the small intestine both directly and indirectly, as undegradable and rumen degradable protein respectively (Alderman and Cottrill, 1993). Also there are positive responses in DM intake to increased concentrations of dietary crude protein, due to beneficial effects on DM digestibility (Oldham, 1984) and improved amino acid supply (Egan, 1965). The Metabolisable Protein System (AFRC, 1992) has attempted to quantify animal protein requirements and protein supply from the diet as metabolisable protein.

Oldham (1984) summarised the responses in total DM intake to protein supplementation as 0.42 kg DM/day for maize silage and 0.19 kg DM/day for grass silage based diets per 10 g/kg DM increase in dietary CP. Newbold (1994) concluded for high yielding dairy cows, that as the ERDP:FME ratio was about 6 g/MJ for maize silage and 14 g/MJ for grass silage (compared with the optimum of 11 g/MJ), intake responses to protein were due to both MP and ERDP supply in the case of maize silage, but in the case of grass silage were due to ERDP supply alone.

The benefits from enhancing the ERDP supply were shown by supplementing maize silage offered as 100% of the forage with 175 g urea/day when the supplement consisted of 6 kg sugarbeet, pulp and 2 kg soyabean meal per day (Leaver, 1991 unpublished). This increased the total diet CP from 133 to 155 g/kg DM. The urea supplementation led to an increase in milk fat content from 431 to 452 g/kg, in milk protein content from 315 to 322 g/kg, and in fat plus protein yield from 1.94 to 2.07 kg/day. These effects were probably a result mainly of the increase in forage DM intake from 13.8 to 14.6 kg DM/day.

In total mixed rations (TMR) with 400 g/kg of the DM as maize silage, increasing the total diet CP content using soyabean meal from 110 to 170 g CP/kg DM led to a substantial increase in total DM intake, in milk yield and in milk composition (Table 9.6). The authors (Kung and Huber, 1983) concluded that combinations of NPN and rumen undegradable protein might be the best way to support protein needs at lowest cost (when supplementing low protein forages). More recently (Broderick, Craig and Ricker, 1993) with TMR diets of 163 g CP/ kg in the total DM, a supplement of soyabean meal plus meat and bone meal was superior to urea (offered at 18 g/kg DM) as the protein supplement especially with a low DM alfalfa silage offered as the companion forage for maize silage (Table 9.7). Nevertheless, the level of urea in this study might have been too high, and this should not preclude urea making a small contribution to rumen degradable protein supply.

**Table 9.6** INFLUENCE OF DIETARY CRUDE PROTEIN CONTENT ON INTAKE AND MILK PRODUCTION IN MAIZE SILAGE BASED DIETS (KUNG AND HUBER, 1983)

| *Diet CP (g/kg DM)* | *110* | *140* | *170* |
|---|---|---|---|
| Total DM intake (kg/day) | 16.5 | 19.4 | 21.4 |
| Milk yield (kg/day) | 31.6 | 33.7 | 34.7 |
| Fat (g/kg) | 29.0 | 30.2 | 30.8 |
| Protein (g/kg) | 28.0 | 29.1 | 29.8 |

Maize silage 400 g/kgDM, alfalfa hay 100 g/kgDM.

The supplementation of urea-treated wholecrop wheat offered as the sole forage with protein has led to some responses in milk production, but not in feed intake (Table 9.8). In Experiment 1, supplementation with soyabean meal increased milk protein yield by 6% when the basal diet was 172 g CP/kg DM. In Experiment 2 with a basal diet of 224 g CP/kgDM there was no significant response. However, fishmeal significantly increased milk yield and milk protein yield in spite of the high CP content of the basal diet. The non-protein nitrogen content of the forages was high in both experiments (273 and 289 g ammonia/kg total N respectively) which might indicate the responses were due to the additional UDP. Nevertheless the estimated MP requirements were more than satisfied by the basal diets.

**Table 9.7** SUPPLEMENTATION OF MAIZE SILAGE/ALFALFA SILAGE USING DIFFERENT PROTEIN SOURCES, AND WITH DIFFERENT DM CONTENTS OF ALFALFA (BRODERICK *et al.* 1993).

| Alfalfa silage | *Low DM* | | *High DM* | |
|---|---|---|---|---|
| Protein supplement | *urea* | *soya bean meal + meat bone meal* | *urea* | *soya bean meal + meat bone meal* |
| Total DM intake (kg/day) | 24.2 | 26.2 | 26.0 | 26.1 |
| Milk yield | 35.4 | 38.5 | 35.9 | 36.9 |
| Fat (g/kg) | 37.0 | 34.8 | 34.7 | 35.4 |
| Protein (g/kg) | 30.9 | 30.1 | 30.4 | 30.2 |

**Table 9.8** SUPPLEMENTATION OF UREA-TREATED WHOLECROP WHEAT WITH PROTEIN (HILL AND LEAVER, 1990; 1992)

| | *Experiment 1* | | *Experiment 2* | | |
|---|---|---|---|---|---|
| CP of concentrate (g/kgDM) | 180 | 240 | 165 | 330 | 330 |
| Source of supplement | - | soya bean meal | - | soya bean meal | fish meal |
| Concentrate (kg DM/day) | 5.2 | 5.2 | 3.5 | 3.5 | 3.5 |
| WC wheat (kg DM/day) | 14.9 | 15.0 | 18.8 | 19.0 | 18.2 |
| Milk yield (kg/day) | 21.3 | 22.0 | 17.3 | 17.2 | 19.0 |
| Fat (g/kg) | 39.2 | 39.5 | 46.1 | 42.7 | 42.8 |
| Protein (g/kg) | 31.7 | 32.4 | 36.8 | 37.3 | 36.6 |
| CP content of total diet (g/kg DM) | 172 | 192 | 224 | 250 | 251 |

A similar conclusion was drawn with growing cattle offered urea-treated wholecrop wheat *ad libitum* (Castejon and Leaver, 1994). Supplementation with fishmeal depressed forage intake, and this was associated with a reduced digestibility of OM and starch (Table 9.9). This indicates that the additional rumen degradable protein supplied by the fishmeal might have been detrimental to rumen conditions in a situation where there was already a massive surplus of ammonia. However, NDF digestibility was increased by both protein supplements which might indicate some enhanced cellulytic activity.

**Table 9.9** EFFECT OF ENERGY/PROTEIN SUPPLEMENTS* ON THE INTAKE AND DIGESTIBILITY OF UREA-TREATED WHOLECROP WHEAT (CASTEJON AND LEAVER, 1994)

|  | *Wholecrop wheat* | *Wholecrop wheat + SBP* | *Wholecrop wheat + SBP + FM* |
|---|---|---|---|
| Supplement intake (kgDM/day) | 0 | 1.6 | 1.6 |
| WC wheat intake (kg DM/day) | 8.6 | 8.7 | 8.2 |
| Digestibility of WC wheat |  |  |  |
| DOMD (g/kg) | 645 | 639 | 626 |
| NDF (g/kg) | 634 | 685 | 675 |
| Starch (g/kg) | 848 | 829 | 815 |

*Mean of two experiments with growing heifers
SBP = molassed sugarbeet pulp. FM = fish meal.

These examples illustrate the complexity of predicting the outcome of supplementation with different levels and sources of protein. One clear conclusion however is that the CP in the total DM should be 160 to 190 g/kgDM, depending on stage of lactation, and that non-protein nitrogen can be used partially, where high proportions of maize and wholecrop silages are in the diet. The observed responses to additional protein up to these levels can be due to a combination of effects; on total DM intake, on MP supply, on body fat mobilisation and on oxidation of MP to enhance energy supply (Newbold, 1994).

## AMOUNT OF CONCENTRATE SUPPLEMENTATION

The ME supplied by an increment of concentrate supplement has a number of influences. It may affect milk yield, milk composition, liveweight change and forage intake. An energy balance will reveal for example where the ME from 1 kg concentrate DM is partitioned. Leaver (1988) showed that on average about one third of the ME from an increment of concentrates was partitioned to milk, one third to liveweight and one third substituted for forage intake. Variations in this partition occur due to stage of lactation, quality of forage, and the basal level of concentrates.

Examples of these responses to increments of concentrates are shown in Tables 9.10 and 9.11 for maize silage and urea-treated wholecrop wheat. The substitution rates in Table 9.10 (Phipps *et al.* 1988) are low, due to a) the poor quality of the grass silage (ME 9.9 MJ/kgDM); and b) the low basal level of concentrates (2.7 kgDM/day). This resulted in high milk yield responses of 1.23 kg milk/kg concentrate DM for grass silage, and 1.08 kg/kg for grass/maize silage mixed. The substitution rate in Table 9.11 for urea-treated wholecrop wheat averaged 0.6 and resulted in an average milk yield response of only 0.5 kg/kg concentrate DM. In both studies the beneficial effects on milk protein content and liveweight change, of incremental increases in concentrates can be clearly seen.

**Table 9.10** LEVEL OF CONCENTRATES (186 G CP/KG DM) FED WITH GRASS SILAGE OR A MIXTURE OF GRASS AND MAIZE SILAGES (PHIPPS *et al.* 1988)

| Forage | Grass silage | | Grass/maize* silage | |
|---|---|---|---|---|
| Concentrate level (kg DM/day) | 2.7 | 5.3 | 2.7 | 5.3 |
| Forage (kg DM/day) | 8.0 | 7.7 | 10.0 | 9.6 |
| Substitution rate | | 0.12 | | 0.15 |
| Milk yield (kg/day) | 16.7 | 19.9 | 20.9 | 23.7 |
| Fat (g/kg) | 37.4 | 38.1 | 38.2 | 36.6 |
| Protein (g/kg) | 27.9 | 29.6 | 29.5 | 30.2 |
| LW change (kg/day) | -0.40 | -0.07 | +0.01 | +0.12 |

*ratio 1:2 (DM basis)

**Table 9.11** LEVEL OF CONCENTRATES (176 G CP/KG DM) FED WITH UREA-TREATED WHOLECROP WHEAT (HILL AND LEAVER, 1990)

| | Concentrate level (kg DM/day) | | |
|---|---|---|---|
| | 5.2 | 7.0 | 8.7 |
| WC wheat intake (kg DM/day) | 14.9 | 13.8 | 12.8 |
| Substitution rate | | 0.61 | 0.59 |
| Milk yield (kg/day) | 21.3 | 22.4 | 23.1 |
| Fat (g/kg) | 39.2 | 40.2 | 39.8 |
| Protein (g/kg) | 31.7 | 32.6 | 33.3 |
| LW change (kg/day) | +0.24 | +0.17 | -0.01 |

The evidence available indicates that the substitution rates of concentrates for forage are little different for maize silage and wholecrop cereals than for grass silage. Therefore substitution rates will in general fall between 0.33 and 0.67, and milk yield responses are likely to be between 0.7 and 1.2 kg/kg concentrate DM. A review of experiments with grass silage as the basal diet (Thomas, 1980), reported a mean short-term response of 0.79 kg milk/kg concentrate DM.

In evaluating the economic returns to increments of concentrate, caution must be taken in estimating cost/benefits which only compare extra milk value against extra concentrate cost. Firstly, the liveweight change resulting from supplementation has a value either directly as beef or indirectly as condition score for the next lactation, and secondly the substitution of concentrates for forage has an economic return in the cost of forage saved. Finally, where milk quota limitations occur, the financial implications of the extra milk production relative to quota costs must also be included in the cost/benefit analysis.

## ENERGY SOURCE IN SUPPLEMENT

The source of energy in the supplement can significantly influence conditions in the rumen especially at high levels of supplementation. At low levels of supplementation eg 5 kg/day of supplement or less, energy source will have little effect on forage intake, milk yield or composition due to its small proportion in the total diet. At higher levels, supplementary energy consisting of starch and sugars will change the balance of microorganisms in the rumen away from cellulytic towards amylolytic activity (Porter, Balch, Coates, Fuller, Latham and Sharpe, 1972; Hoover, 1986). This will reduce the rate and extent of fibre digestion (Terry, Tilley and Outen, 1969) with consequential effects on forage intake (Thomas, 1987).

There has been a large number of experiments to examine starch versus fibre-based concentrates with grass silage and hay diets (eg Mayne and Gordon, 1984; Sloan, Rawlinson and Armstrong, 1988; Sutton, Morant, Bines, Napper and Givens, 1993; Aston, Thomas, Daley and Sutton, 1994). Starch supplements have invariably led to a reduction in milk fat content, and to an increase in milk protein content and/or yield. The reduced milk fat content is caused by a higher proportion of propionic acid being produced in the rumen at the expense of acetic acid which is a precursor of milk fat, and is a result of the greater amylolytic activity of the rumen microbes (Thomas and Chamberlain, 1984; De Visser, 1993). The positive responses of milk protein to starch compared with fibre-based concentrates are partly associated with the beneficial effects on microbial protein synthesis (as FME supply), and partly to the supply of additional propionic acid which reduces the necessity for the use of amino acids as glucose precursors, particularly during negative energy balance (De Visser, 1993).

Very few comparisons have been made of starch versus fibre-based concentrates with maize silage or wholecrop cereal forages. Kristensen (1992) compared rolled barley and dried sugarbeet pulp as supplements for ensiled wholecrop barley, offered as the sole forage (Table 9.12). Wholecrop barley intake increased when rolled barley was replaced by sugarbeet pulp, milk fat content increased and milk protein content declined. These results are similar to those reported for grass silages. However, immature wholecrop barley has some similarities to grass silage so this was possibly to be expected.

**Table 9.12** ENERGY SOURCE OF SUPPLEMENT FED WITH ENSILED WHOLECROP BARLEY (KRISTENSEN, 1992)

|                          | *RB* | *RB/SBP* | *SBP* |
|--------------------------|------|----------|-------|
| Concentrate (kg DM/day)  | 8.2  | 8.2      | 8.2   |
| WC barley (kg DM/day)    | 9.0  | 10.6     | 10.0  |
| Milk yield (kg/day)      | 28.1 | 27.6     | 27.3  |
| Fat (g/kg)               | 40.0 | 41.6     | 41.9  |
| Protein (g/kg)           | 31.7 | 31.6     | 30.0  |

RB = rolled barley   SBP = molassed sugarbeet pulp

Maize silage and more mature wholecrop cereals have a higher grain content than ensiled wholecrop wheat or barley, and some differences in response to starch versus fibre supplement might be expected. The results for urea- treated wholecrop wheat shown in Table 9.13 indicated no significant differences in milk yield or composition between a high fibre and high starch plus sugar concentrate. Therefore, the source of energy in the supplement may not be influential at a moderate level of alternative forage inclusion (400 g/kg of forage DM) and at a moderate concentrate input (8 kg/day).

**Table 9.13** ENERGY SOURCE OF SUPPLEMENT FED WITH UREA-TREATED WHOLECROP WHEAT AND GRASS SILAGE (LEAVER AND MARSDEN, 1991 UNPUBLISHED)

| *Supplement**            | *High fibre* | *High starch + sugar* |
|--------------------------|--------------|-----------------------|
| Concentrate (kg DM/day)  | 6.9          | 6.9                   |
| Forage (kg DM/day)+      | 11.6         | 12.0                  |
| Milk yield (kg/day)      | 29.6         | 29.9                  |
| Fat (g/kg)               | 39.8         | 40.1                  |
| Protein (g/kg)           | 31.2         | 31.6                  |

* High fibre: 277 g starch + sugar/kgDM and 142 g crude fibre/kg DM
  High starch + sugar: 376 g starch + sugar/kg DM and 76 g crude fibre/kg DM
+ 60:40 ratio of grass silage : urea-treated wholecrop wheat

There is a developing interest in different types of starch as supplements for forages. It is well established that maize starch has a low rumen degradability compared with wheat or barley starch (De Visser, 1993), and therefore there are implications for fibre digestion, forage intake and microbial protein production. In a recent study (Overton *et al.* 1995) rolled barley replaced shelled corn (maize grain) as the energy supplement (Table 9.14). This replacement led to a reduction in total DM intake, a reduction in milk yield, and an increase in milk fat and protein contents. However the yields of milk fat and protein declined. A higher proportion of starch was digested in the rumen with barley than with corn supplements, and this was associated with a decrease in acetic and an increase in propionic acid production. The higher DM intake of the shelled corn diets was the explanation for the higher milk yields and for the dilution effect on milk solids content.

**Table 9.14** SHELLED MAIZE AND STEAM ROLLED BARLEY AS SUPPLEMENTS FOR MAIZE/ALFALFA SILAGE (OVERTON *ET AL.* 1995)

|  | Ratio corn : barley | | |
| --- | --- | --- | --- |
|  | *100 : 0* | *50 : 50* | *0 : 100* |
| Total intake (kg DM/day) | 22.8 | 21.3 | 19.6 |
| % starch intake digested in rumen | 42 | 61 | 74 |
| % starch intake digested post ruminally | 49 | 33 | 22 |
| % NDF intake digested in rumen | 40 | 43 | 35 |
| % NDF intake digested post ruminally | 11 | 3 | 12 |
| Milk yield (kg/day) | 26.9 | 26.6 | 22.6 |
| Fat (g/kg) | 35.8 | 35.0 | 39.1 |
| Protein (g/kg) | 33.6 | 34.4 | 36.9 |

TMR with CP 162 g/kgDM, NDF 346 g/kgDM, starch 329 g/kg DM, and forage 400g/kg of total DM.

The processing of grain (physical and chemical) can also be influential on milk production through its effect on diet digestibility, rate and site of grain digestion and on food intake (Campling, 1991). Processing which leads to an increase in access of rumen microorganisms to the starch, will increase both the rate and extent of digestion in the rumen, particularly for maize grain which is most resistant to microbial attack. Nonetheless the quantification of the effects of processing are not clear due to interactions with the forage and other components of the diet (Owens, Zinn and Kim, 1986; Theurer, 1986). For dairy cattle some processing of cereal grains is recommended except perhaps for oats which can be fed whole (Orskov, 1987; Campling, 1991).

There is no information available on the use of dietary fats specifically as supplements for maize silage or wholecrop cereals. Thomas and Martin (1988) reviewed the role of supplemental fats, and concluded that with 'fat deficient' diets, supplementation leads to increases in milk yield and milk composition, especially in milk fat content. At the other extreme (over 100 g fat/kg total DM) interference of fat in rumen fermentation leads to reduced feed intake and depressed milk yield and composition. For diets between these extremes a range of effects of fat supplementation have been recorded depending on fat source and level, and the type of basal diet (Storry and Brumby, 1980; Palmquist, 1984). Whilst the response in milk fat content to moderate levels of fat supplementation is variable, responses in milk yield are generally positive and in milk protein, negative, even with 'protected fat' products (Thomas and Martin, 1988).

## Optimising supplementation

The term "optimising" presents a challenge to whoever is carrying out the task, as it poses the question, optimising for what or whom? Decisions concerning the supplementary feeding of dairy cows offered maize silage or wholecrop cereals could be aimed at optimising rumen conditions for microbial growth, optimising nutrient supply to the cow for reproduction and health, optimising to meet milk quota targets, optimising for profit for the farmer and many more. These objectives would not all require the same type and level of supplement input. Therefore quantification of these relationships is a priority in order to provide the necessary information for nutritional decision making at rumen, cow and farm level.

### SUPPLEMENTATION OF THE RUMEN

The development of the MP system (AFRC, 1992) focused attention on the rumen and microbial protein production, and the system has been integrated with the ME system into a practical manual for farmers and advisors (Alderman and Cottrill, 1993). Arising from this, and similar developments in other countries, ruminant nutritionists have become increasingly interested in synchrony between energy and N supply for the rumen microorganisms (Herrera-Saldana, Gomez-Alarcon, Torabi and Huber, 1990; Sinclair, Garnsworthy, Newbold and Buttery, 1993). Newbold (1994) has defined rumen synchrony as 'ensuring the optimal ratio between N and energy supply throughout 24 hours of the day.' He highlighted the problems of confounding differences in rate of degradation with extent of degradation in the rumen, which has occurred in a number of experiments examining synchrony.

The concept of rumen synchrony is potentially useful as an objective for nutritional management. However, with the present state of knowledge, providing advice on optimisation of synchrony is difficult. Although a body of information is being built up on rates and extent of degradation of separate feeds, the outcome of their interactions in the rumen for animals of different production (and therefore intake) levels and with different diurnal feeding patterns is extremely difficult to predict. Nevertheless this is a fruitful area for research.

## SUPPLEMENTATION OF THE COW

The objectives of supplementary feeding the cow are to influence milk yield and composition, although the effects on reproduction and health, if known, must also be taken into account.

The effects of amount of concentrates, protein source and level, and energy source have been discussed earlier. Decision making at farm level is normally through ration-formulation procedures which attempt to match nutrient supply to nutrient requirements for individuals or groups of cows. The ME and MP systems have been developed for this purpose, and are more useful in attempting to match diet to cow requirements than to predict responses to changes in diet (Newbold, 1994). This particularly applies to protein where a change in input may affect DM intake as well as the supply of MP.

Systems of feeding are increasingly based on group rather than individual cow concentrate allocation especially through TMR feeding or flat-rate feeding in the milking parlour. TMR feeding should have advantages in rumen synchrony and this might explain why the system has in some instances increased feed intake and milk composition (Phipps, Bines, Fulford and Weller, 1984; Nocek, Steele and Braund, 1986). Nevertheless comparisons of mixed diets with forages and concentrates fed separately have in general not produced clear effects on milk production but where large quantities of concentrates are fed, TMR or out-of-parlour feeding systems become a necessity. Infrequent feeding of large amounts of concentrates gives asynchrony in the rumen by reducing rumen pH, which in turn reduces feed intake and this may reduce milk production especially milk fat content and fat yield (Gibson, 1984).

The effects of supplementation on reproduction and health are complex and not well understood. In early lactation, concentrate supplementation may alleviate the mobilisation of body tissues and this may be beneficial to reproductive efficiency (Garnsworthy and Haresign, 1989). Evidence from the USA indicates that high dietary protein levels (190 to 210 g CP/kg DM) can lead to reduced conception rates, an increased incidence of cystic ovaries, endometritis and dystocia (Carroll, Barton, Anderson and Smith, 1988). The reduced conception rates could be due to an enhanced body tissue mobilisation or to high levels of blood urea. Ferguson and Chalupa (1989) have indicated that the effects are mainly accounted for by rumen degradable protein intake. If this is confirmed, the high levels of ammonia in urea-treated wholecrop cereals, if fed as a large proportion of the diet, could be detrimental to fertility. Also, the level and degradability of the protein supplement offered in early lactation could be influential on fertility.

The prevalence of lameness in herds can be affected by the source and level of supplementation (Leaver, 1990), although this has not been shown with maize silage or wholecrop cereal diets. High protein diets in early lactation (196 g CP/ kg DM) have been shown to increase the prevalence of lameness compared with lower protein diets (160 g CP/kg DM). This appeared to be due to faster and softer hoof growth, and which led to adverse effects on locomotion and clinical lameness (Manson and Leaver, 1988). High starch compared with high fibre concentrates also increase the prevalence of lameness due to an increase in the incidence of solar ulcers resulting from laminitis and poor quality horn growth (Kelly and Leaver, 1990). Avoidance of extreme diets in CP content, and starch content may therefore be beneficial to both reproduction and lameness.

Targets for diet formulation change with time, partly due to new knowledge arising and partly to fashion. The present trend is towards USA targets for total diet specification, such as those shown in Table 9.15. There is a need to test the validity of such targets in animal response experiments and to determine cost/ benefits of alternative approaches. For example the ration shown in Table 9.16 has a diet specification very different from those in Table 9.15, yet for a farmer with ample quantities of forage available, this may be a perfectly satisfactory nutritional solution.

**Table 9.15** RATION SPECIFICATIONS FOR LACTATING DAIRY COWS (POND *et al.* 1995)

|  | Stage of lactation | | |
|---|---|---|---|
|  | *Early* | *Mid* | *Late* |
| DM intake (kg/100 kg LW) | > 4.0 | 3.5-4.0 | 3.0-3.5 |
| (g/kg DM) | | | |
| CP | 170-180 | 160-170 | 140-160 |
| SIP | 300-350 | 350-400 | 350-400 |
| UIP | 350-400 | 350-400 | 350-400 |
| NDF | 260-300 | 320-340 | 340-360 |
| ADF | 180-200 | 210-230 | 220-240 |
| NSC | 350-400 | 350-400 | 350-400 |
| Fat (maximum) | 60-80 | 40-60 | 40-50 |

SIP = soluble protein. UIP = undegradable protein. NSC = non-structural carbohydrate.

**Table 9.16** HIGH FORAGE DIET FOR LACTATING DAIRY COWS* (LEAVER, 1993 UNPUBLISHED)

| Diet | | Diet specification (kg DM/day) | | Performance (g/kg DM) | |
|---|---|---|---|---|---|
| Soyabean meal | 1.8 | CP | 160 | Milk yield (kg/day) | 27.4 |
| Rapeseed meal | 1.8 | NDF | 438 | Fat (g/kg) | 43.7 |
| Grass silage | 4.9 | ADF | 271 | Protein (g/kg) | 33.1 |
| Maize silage | 11.4 | NSC | 171 | LW gain (kg/day) | 0.31 |
| Total (kg/100 kg LW) | 3.29 | | | | |

*at week 20 of lactation

## SUPPLEMENTATION FOR FARM PROFITABILITY

Nutritional advice to the farmer, whilst taking into account the implications for rumen microorganisms, and cow requirements, must be based on sound economics. What is optimal for the rumen microorganisms or for the individual cow, may not be optimal for farm profit. The resources of land, labour, capital and milk quota differ for each farm and therefore nutritional advice has to be tailored to each farm circumstance.

The optimum supplementation of maize silage and wholecrop cereal based diets will therefore be determined by a range of factors including the amount and quality of forages available and milk quota targets, as well as the needs of the rumen microorganisms and the potential of the cows.

## Conclusions

Maize silage and wholecrop cereals are replacing grass silage in the diet of dairy cows in many areas of the country. This trend seems likely to continue, but due to the complementary benefits between these forages (Phipps, 1994), inclusion rates (proportion of total forage) with grass silage will generally range from 25 to 75%.

These forage alternatives to grass silage are low in CP, minerals and vitamins and appropriate supplementation is required. Increasing the CP of the total diet to 160 to 190 g CP/kg DM (level depending on stage of lactation) increases DM intake, milk yield and milk constituent yield. Rumen degradable protein can contribute partially to the protein supplementation.

Supplementation with concentrates appears to give similar substitution rates to grass silage (generally 0.33 to 0.67) and therefore responses in milk yield are likely to be similar (0.7 to 1.2 kg milk/kg concentrate DM).

Maize starch has a lower degradability than wheat or barley starch, and this has implications for maize and wholecrop cereal forages as well as for types of cereal supplements. There are implications for FME supply, VFA proportions in the rumen, microbial protein production, feed intake and milk production. Future nutritional research will increasingly focus on attempting to synchronise the release of energy and protein substrates in the rumen to optimise rumen conditions for microbial growth.

For high genetic merit cows there is a trend towards diets which include maize silage or wholecrop cereals in which supplementation is targeted at producing much lower total diet NDF and higher NSC levels, than used previously. Whilst the objectives of improving rumen synchrony and meeting the requirements of high potential cows are logical nutritionally, an important objective of the farmer is to feed the herd profitably. There may be times when the three objectives are in conflict.

# References

AFRC (1992) Technical Committee on Responses to Nutrients. Report No. 9. Nutritive Responses of Ruminant Animals: Protein. *Nutrition Abstracts and Reviews (Series B)*, **62**, 787-835.

Alderman, G. and Cottrill, B.R. (1993) *Energy and Protein Requirements of Ruminants*. Wallingford: CAB International.

Aston, K., Thomas, C., Daley, S.R. and Sutton, J.D. (1994) Milk production from grass silage diets: effects of the composition of supplementary concentrates. *Animal Production*, **59**, 335-344.

Broderick, G.A., Craig, W.M. and Ricker, D.B. (1993) Urea vs true protein as a supplement for lactating dairy cows fed grain plus mixtures of alfalfa and corn silages. *Journal of Dairy Science*, **76**, 2266-2274.

Campling, R.C. (1991) Processing cereal grains for cattle - a review. *Livestock Production Science*, **28**, 223-234.

Carroll, D.J., Barton, B.A., Anderson, G.W. and Smith, R.D. (1988) Influence of protein intake and feeding strategy on reproductive performance of dairy cows. *Journal of Dairy Science*, **71**, 3470-3481.

Castejon, M. and Leaver, J.D. (1994) Intake and digestibility of urea-treated wholecrop wheat and liveweight gain by dairy heifers. *Animal Feed Science and Technology*, **46**, 119-130.

De Visser, H. (1993) Characterisation of carbohydrates in concentrates for dairy cows. In *Recent Advances in Animal Nutrition - 1993*, pp 19-38. Edited by P.C. Garnsworthy and D.J.A. Cole. Nottingham: Nottingham University Press.

Egan, A.R. (1965) Nutritional status and intake regulation in sheep. II The influence of sustained duodenal infusions of casein or urea upon voluntary intake of low-protein roughages by sheep. *Australian Journal of Agricultural Research*, **16**, 451-462.

Ferguson, J.D. and Chalupa, W. (1989) Impact of protein nutrition on reproduction in dairy cows. *Journal of Dairy Science*, **72**, 746-766.

Garnsworthy, P.C. and Haresign, W. (1989) Fertility and nutrition. In *Dairy Cow Nutrition - The Veterinary Angles*, pp 23-34. Edited by A.T. Chamberlain. Reading: University of Reading.

Gibson, J.P. (1984) The effects of feeding frequency on milk production of dairy cattle: an analysis of published results. *Animal Production*, **38**, 181-189.

Givens, D.I., Moss, A.R. and Adamson, A.H. (1993) The digestion and energy value of wholecrop wheat treated with urea. *Animal Feed Science and Technology*, **43**, 51-64.

Givens, D.I., Cottyn, B.G., Dewey, P.J.S. and Steg, A. (1995). A comparison of the neutral detergent-cellulase method with other laboratory methods for predicting the digestibility *in vivo* of maize silages from three European countries. *Animal Feed Science and Technology*, **54**, 55-64.

Hargreaves, A. (1993) *Wholecrop barley silage as a supplementary feed for grazing dairy cows*. PhD thesis: Wye College, University of London.

Herrera-Saldana, R., Gomez-Alarcon, R., Torabi, M. and Huber, J.T. (1990) Influence of synchronizing protein and starch degradation in the rumen on nutrient utilisation and microbial protein synthesis. *Journal of Dairy Science*, **73**, 142-148.

Hill, J. and Leaver, J.D. (1990) Urea-treated wholecrop wheat for dairy cattle. *Animal Production*, **50**, 578.

Hill, J. and Leaver, J.D. (1992) Effect of protein supplementation on the digestibility of urea-treated wholecrop wheat and on milk production by dairy cows. *Animal Production*, **54**, 449-450.

Hill, J. and Leaver, J.D. (1993) The intake, digestibility and rate of passage of wholecrop wheat and grass silage by growing heifers. *Animal Production*, **56**, 443.

Hoover, W.H. (1986) Chemical factors involved in ruminal fiber digestion. *Journal of Dairy Science*, **69**, 2755-2766.

Kelly, E.F. and Leaver, J.D. (1990) Lameness in dairy cattle and the type of concentrate given. *Animal Production*, **51**, 221-227.

Kristensen, V.F. (1992) The production and feeding of whole-crop cereals and legumes in Denmark. In *Whole-Crop Cereals*, pp 21-37. Edited by B.A. Stark and J.M. Wilkinson. Canterbury: Chalcombe Publications.

Kung, L. and Huber, J.T. (1983) Performance of high producing cows in early lactation fed protein of varying amounts, sources, and degradability. *Journal of Dairy Science*, **66**, 227-234.

Kung, L., Tung, R.S. and Carmean, B.R. (1992) Rumen fermentation and nutrient digestion in cattle fed diets varying in forage and energy source. *Animal Feed Science and Technology*, **39**, 1-12.

Leaver, J.D. (1988) Level and pattern of concentrate allocation to dairy cows. In *Nutrition and Lactation in the Dairy Cow*, pp 315-326. Edited by P.C. Garnsworthy. London: Butterworth.

Leaver, J.D. (1990) Effects of feed changes around calving on cattle lameness. In *Update in Cattle Lameness, Proceedings of the VIth International Symposium on Diseases of the Ruminant Digit*, pp 102-108. Neston: The British Cattle Veterinary Association.

Leaver, J.D. (1991) *Forage maize production for dairy cattle. A Farm Study*. Wye: Wye College.

Leaver, J.D. and Hill, J. (1992) Feeding cattle on whole-crop cereals. In *Whole-Crop Cereals*, pp 59-69. Edited by B.A. Stark and J.M. Wilkinson. Canterbury: Chalcombe Publications.

Leaver, J.D. and Hill, J. (1995) The performance of dairy cows offered ensiled whole-crop wheat, urea-treated whole-crop wheat or sodium hydroxide-treated wheat grain and wheat straw in a mixture with grass silage. *Animal Science*, **61**, 481-489.

Manson, F.J. and Leaver, J.D. (1988) The influence of dietary protein intake and of hoof trimming on lameness in dairy cattle. *Animal Production*, **47**, 191-199.

Mayne, C.S. and Gordon, F.J. (1984) The effect of type of concentrate and level of concentrate feeding on milk production. *Animal Production*, **39**, 65-76.

Newbold, J.R. (1994) Practical application of the metabolisable protein system. In *Recent Advances in Animal Nutrition - 1994*, pp 231-264. Edited by P.C. Garnsworthy and D.J.A. Cole. Nottingham: Nottingham University Press.

Nocek, J.E., Steele, R.L. and Braund, D.G. (1986) Performance of dairy cows fed forage and grain separately versus a total mixed ration. *Journal of Dairy Science*, **69**, 2140-2147.

Oldham, J.D. (1984) Protein-energy interrelationships in dairy cows. *Journal of Dairy Science*, **67**, 1090-1114.

Orskov, E.R. (1987) *The Feeding of Ruminants*. Marlow: Chalcombe Publications.

Overton, T.R., Cameron, M.R., Elliott, J.P., Clark, J.H. and Nelson, D.R. (1995) Ruminal fermentation and passage of nutrients to the duodenum of lactating cows fed mixtures of corn and barley. *Journal of Dairy Science*, **78**, 1981-1998.

Owens, F.N., Zinn, R.A. and Kim, Y.K. (1986) Limits to starch digestion in the ruminant small intestine. *Journal of Animal Science*, **63**, 1634-1648.

Palmquist, D. (1984) Use of fats in diets for lactating dairy cows. In *Fats in Animal Nutrition*, pp 357-381. Edited by J. Wiseman. London: Butterworths.

Phipps, R.H. (1994) Complementary forages for milk production. In *Recent Advances in Animal Nutrition - 1994*, pp 215-230. Edited by P.C. Garnsworthy and D.J.A. Cole. Nottingham: Nottingham University Press.

Phipps, R.H., Bines, J.A., Fulford, R.J. and Weller, R.F. (1984) Complete diets for dairy cows: a comparison between complete diets and separate ingredients. *Journal of Agricultural Science, Cambridge*, **103**, 171-180.

Phipps, R.H., Weller, R.F., Elliot, R.J. and Sutton, J.D. (1988) The effect of level and type of concentrate and type of conserved forage on dry matter intake and milk production of lactating dairy cows. *Journal of Agricultural Science, Cambridge*, **111**, 179-186.

Phipps, R.H., Sutton, J.D. and Jones, B.A. (1995) Forage mixtures for dairy cows: the effect on dry-matter intake and milk production of incorporating either fermented or urea-treated whole-crop wheat, brewers' grains, fodder beet or maize silage into diets based on grass silage. *Animal Science*, **61**, 491-496.

Pond, W.G., Church, D.C. and Pond, K.R. (1995) *Basic Animal Nutrition and Feeding. Fourth Edition.* New York: John Wiley and Sons.

Porter, J.W.G., Balch, C.C., Coates, M.E., Fuller, R., Latham, M.J. and Sharpe, M. (1972) The influence of gut flora on the digestion, absorption and metabolism of nutrients in animals. *Biennial Reviews*, pp 13-36. Reading: National Institute for Research in Dairying.

Sinclair, L.A., Garnsworthy, P.C., Newbold, J.R. and Buttery, P.J. (1993) Effect of synchronizing the rate of dietary energy and nitrogen release on rumen fermentation and microbial protein synthesis in sheep. *Journal of Agricultural Science, Cambridge*, **120**, 251-264.

Sloan, B.K., Rowlinson, P. and Armstrong, D.G. (1988) Milk production in early lactation by dairy cows given grass silage *ad libitum*: influence of concentrate energy source, crude protein content and level of concentrate allowance. *Animal Production*, **46**, 317-331.

Storry, J.E. and Brumby, P.E. (1980) Influence of nutritional factors on the yield and content of milk fat: protected non-polyunsaturated fat in the diet. In *Factors Affecting the Yields and Contents of Milk Constituents of Commercial Importance*, pp 105-120. International Dairy Federation, document no 125, Brussels: IDF.

Sutton, J.D., Morant, S.V., Bines, J.A., Napper, D.J. and Givens, D.I. (1993) Effect of altering the starch:fibre ratio in the concentrates on hay intake and milk production by Friesian cows. *Journal of Agricultural Science, Cambridge*, **120**, 379-390.

Sutton, J.D., Abdalla, A.L., Phipps, R.H., Cammell, S.B. and Humphries, D.J. (1995) Digestibility and nitrogen balance in dairy cows given diets of grass silage and wholecrop wheat. *Animal Science*, **60**, 510.

Terry, R.A., Tilley, J.M.A. and Outen, G.E. (1969) Effect of pH on the cellulose digestion under *in vitro* conditions. *Journal of the Science of Food and Agriculture*, **20**, 317-320.

Theurer, C.B. (1986) Grain processing effects on starch utilization by ruminants. *Journal of Animal Science*, **63**, 1649-1662.

Thomas, C. (1980) Conserved forages. In *Feeding Strategies for Dairy Cows*, pp 8.1-8.14. Edited by W.H. Broster, C.L. Johnson and J.C. Tayler. London: Agricultural Research Council.

Thomas, C. (1987) Factors affecting substitution rates in dairy cows on silage-based rations. In *Recent Advances in Animal Nutrition - 1987*, 205-218. Edited by W. Haresign. London: Butterworths.

Thomas, P.C. and Chamberlain, D.G. (1984) Manipulation of milk composition to meet market needs. In *Recent Advances in Animal Nutrition - 1984*, pp 219-245. Edited by W. Haresign and D.J.A. Cole. London: Butterworths.

Thomas, P.C. and Martin, P.A. (1988) The influence of nutrient balance on milk yield and composition. In *Nutrition and Lactation in the Dairy Cow*, pp 97-118. Edited by P.C. Garnsworthy. London: Butterworths.

Wilkinson, J.M. (1978) The ensiling of forage maize: effects on composition and nutritive value. In *Forage Maize*, pp 201-237. Edited by E.S. Bunting, B.F. Pain, R.H. Phipps, J.M. Wilkinson and R.E. Gunn. London: Agricultural Research Council.

**V**

**Pig Nutrition**

**10**

# IMPACT OF IMMUNE SYSTEM ACTIVATION ON GROWTH AND OPTIMAL DIETARY REGIMENS OF PIGS

TIM STAHLY, PH.D.

*Professor of Animal Science, Iowa State University, Ames, Iowa 50011, USA*

Exposure of animals to substances foreign to the body can result in severe alteration in normal body function up to the point that death occurs. These substances, called antigens, include bacteria, viruses, pesticides, and soya proteins. Fortunately, animals possess body defence systems (i.e. immune system) that function to contain or destroy such antigens before life-threatening changes occur. The general response of the immune system to an antigen is initiated by a release of a series of compounds called cytokines. Release of specific cytokines activates the cellular (i.e. phagocytic) and humoral (i.e. antibody) components of the immune system. Cytokine release also alters various endocrine pathways and metabolic processes in the body. ACTH and thyroxine, which are catabolic hormones, are elevated (Navarra *et al.*, 1991; Hashimoto et al., 1994). Thymic and somatotropic peptides, which are potential anabolic hormones, are inhibited (Hannager *et al.*, 1991; Fan et al., 1994). Voluntary feed intake is lowered (Mrosovsky *et al.*, 1989; Johnson and von Borell, 1994). Core body temperature and body heat production are increased. Overall, body protein synthesis is reduced, and more body proteins are degraded as part of the body's defence to fight the invading antigens (Klasing *et al.*, 1987; Ballmer *et al.*, 1991; Zamir *et al.*, 1994). These metabolic adjustments result in a reduced rate of body growth, less efficient utilization of feed for growth and potentially fatter carcasses.

## Impact of chronic immune system activation on body growth

At our research station, we have initiated studies to determine the impact of minimizing the activation of the pig's immune system on the rate, efficiency and composition of growth as well as the optimum dietary regimens of pigs. Two management schemes for the sow and pig have been utilized to achieve animals with a low and high level of chronic immune system (IS) activation. A medicated early-weaning scheme has been utilized to create animals with a low level of immune activation. In this scheme, the pigs' dams are vaccinated for the prevalent antigens (i.e. *Mycoplasma hyopneumonia, Actinobacillus pleuropneumoniae*) in the herd and the neonatal pigs receive injectable antimicrobial agents effective against the prevalent antigens at d 1, 3, 5, 8 and 11 of age. The pigs are weaned at 10 to 14 days of age into an off-site nursery isolated from other pigs. A conventional weaning scheme has been utilized to create animals with a high level of immune activation. The dams of these pigs are not vaccinated nor do the pigs receive antimicrobial agents preweaning. The pigs are weaned at 19 to 21 days of age into an on-site nursery previously occupied by pigs. Sows and pigs utilized in both schemes are derived from the same genetic line and geographic site of origin. In each immune status group, pigs receive milk diets to day 19 of age.

In the initial studies, pigs in each immune status group were allowed to consume a fortified corn-soyabean meal, 20% dried whey, 5% dried skim milk mixture *ad libitum* from 19 days of age until the pigs reached body weights of about 27 kilograms. Immune parameters were monitored to verify that a low and high level of chronic IS activation was maintained during the study. The pigs in the low IS possessed fewer CD4 positive T-lymphocytes, more CD8 positive T-lymphocytes, and lower CD4:CD8 ratios than the high IS pigs at 9, 17, and 25 kg body weight. The low IS pigs also had lower serum concentration of alpha-l-acylglycoprotein (an acute phase protein) at 9, 17, and 25 kg body weight. Pigs in the low IS group consumed more feed, grew faster and required less feed per unit of gain than those in the high IS group (Table 10.1). Furthermore, pigs in the low IS group deposited more proteinaceous body tissue and produced carcasses with less body fat relative to proteinaceous tissue.

In subsequent studies, the impact of minimizing the activation of the pig's immune system on growth, particularly muscle growth, from 6 to 113 kilograms body weight has been quantified. Again, minimizing antigen exposure and thus immune system activation throughout the pig's development resulted in higher feed intakes, greater body weight gains, more body muscle development and less feed required per kg of growth (Table 10.2).

The immune status of the animals also influences the milk yield capacity of mammary tissue. Systemic and intramammary infusion of endotoxin in dairy cows result in depressed milk yields (Shuster *et al.*, 1991a,b). The depressed milk yield is assumed to be due in part to the impact of cytokines on milk protein synthesis. It is hypothesized that a similar reduction in milk yield would occur in sows if systemic release of cytokines occurred; however, this response has not been evaluated.

**TABLE 10.1** IMPACT OF LEVEL OF CHRONIC IMMUNE SYSTEM (IS) ACTIVATION ON RATE, EFFICIENCY AND COMPOSITION OF GROWTH IN PIGS FED FROM 6 TO 27 KILOGRAMS BODYWEIGHT[A]

| | IS activation | | Unit |
| Item | Low | High | Change |
| --- | --- | --- | --- |
| Pig body weight, kg | | | |
| Initial | 6.4 | 5.9 | |
| Final | 27.2 | 25.9 | |
| Growth and feed utilization | | | |
| Daily feed, g | 973 | 863 | +110 |
| Daily gain, g | 676 | 477 | +199 |
| Feed/gain | 1.44 | 1.81 | -.37 |
| Composition of body growth[b] | | | |
| Protein gain, g/day | 105 | 65 | +40 |
| Fat gain, g/day | 67 | 63 | +4 |
| Fat to protein gain | .64 | .95 | -.31 |

[a]Adapted from Williams *et al.* (1993a,b). Pigs (castrates only) allowed to consume diets *ad libitum*.
[b]Body nutrient accretion determined using a comparative slaughter technique.

**Table 10.2** IMPACT OF LEVEL OF CHRONIC IMMUNE SYSTEM (IS) ACTIVATION ON RATE, EFFICIENCY AND COMPOSITION OF GROWTH IN PIGS FED FROM 6 TO 113 KILOGRAMS BODY WEIGHT[a]

| | IS activation | | Unit |
| Item | Low | High | Change |
| --- | --- | --- | --- |
| Growth and feed utilization | | | |
| Daily feed, g | 2,296 | 2,066 | +230 |
| Daily gain, g | 850 | 677 | +173 |
| Feed/gain | 2.70 | 3.05 | -.35 |
| Carcass traits | | | |
| Backfat tenth rib, mm | 27.6 | 31.4 | -3.8 |
| L. muscle area, cm$^2$ | 37.4 | 32.6 | +4.8 |
| Carcass muscle, % | 55.8 | 52.5 | +3.3 |

[a] Adapted from Williams *et al.* (1994) and Stahly *et al.* (1994b). Pigs (castrates only) allowed to consume diets *ad libitum*.

## Impact of immune system activation on nutrient needs

The intakes of certain dietary nutrients, particularly amino acids, need to be increased to support the elevated rates of the body growth in the low immune status pigs. In two of the studies outlined above, pigs in each immune status group were allotted to one of five dietary amino acid regimes. From 6 to 27 kilograms body weight, diets formulated to contain 6, 9, 12, 15 and 18 g lysine/kg diet were fed. From 27 to 111 kilograms, these lysine concentrations were reduced to 4.5, 6, 7.5, 9 and 10.5 g lysine/kg diet, respectively. The dietary lysine concentrations were achieved by altering the ratio of corn and soybean meal in the diets.

The growth responses of the two immune status groups to the five dietary amino acid regimens fed from 6 to 27 kilograms body weight are shown in Table 10.3. Pigs experiencing the low level of IS activation consumed more feed-nutrients daily, gained body weight faster and required less feed per unit of weight gain. Furthermore, a higher proportion of their body weight gain was protein and a lower proportion was fat. Because of their greater capacity for proteinaceous tissue accretion, the low IS pigs needed a greater dietary intake of the limiting amino acid (i.e., lysine) expressed as grams per day or concentration in the diet to maximize rate and efficiency of growth than the high IS pigs.

**Table 10.3**  IMPACT OF LEVEL OF CHRONIC IMMUNE SYSTEM (IS) ACTIVATION AND DIETARY AMINO ACID CONCENTRATION ON GROWTH OF PIGS FED FROM 6 TO 27 KILOGRAMS BODY WEIGHT[a]

| Item | IS Activation | *Dietary lysine, g/kg diet*[b] | | | | |
| | | 6 | 9 | 12 | 15 | 18 |
|---|---|---|---|---|---|---|
| **Feed-nutrient intake** | | | | | | |
| Daily feed, g | Low | 1114 | 1052 | 990 | 977 | 968 |
| | High | 870 | 927 | 907 | 863 | 883 |
| Daily lysine, g | Low | 6.7 | 9.5 | 11.9 | 14.6 | 17.4 |
| | High | 5.2 | 8.3 | 10.9 | 12.9 | 15.9 |
| **Growth and feed utilization** | | | | | | |
| Daily gain, g | Low | 476 | 577 | 652 | 677 | 624 |
| | High | 364 | 478 | 531 | 476 | 491 |
| Feed/gain | Low | 2.35 | 1.82 | 1.52 | 1.44 | 1.55 |
| | High | 2.39 | 1.93 | 1.70 | 1.82 | 1.80 |

[a]Adapted from Williams *et al.* (1994). Pigs allowed to consume corn-soyabean meal, 20% dried whey, 5% dried skim milk mixture fortified with minerals and vitamins *ad libitum* from 6 to 27 kg body weight.
[b]Dietary lysine concentrations achieved by altering the ratio of corn and soyabean meal in the diet.

Over the duration of this study, pig weights and feed consumption were determined at 7-day intervals to determine the optimum dietary lysine concentration at each stage of the pig's development. The dietary lysine concentrations needed to optimize efficiency of feed utilization in the two immune status groups at each stage of development from 9 to 109 kilograms are outlined in Figure 10.1. The pigs in the low IS group needed a higher dietary concentration as well as a daily intake of the limiting amino acid at each stage of growth evaluated.

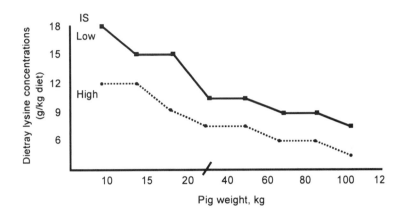

**Figure 10.1** Dietary lysine concentrations needed to optimize efficiency of feed utilization for specific increments of body development in pigs with a low or high level immune system (IS) activation. Data derived from castrates with a moderate genetic capacity for lean tissue growth that are housed in a thermoneutral climate and allowed to consume a corn-soyabean meal, 20% whey, 5% skim milk diet and a corn-soyabean meal diet ad libitum from 6 to 27 and 27 to 111 kilogram body weight, respectively. Adapted from Williams *et al.* (1994).

Because the amino acid makeup of proteins in tissues differs from those associated with maintenance functions, dietary amino acid needs are not altered equally when the animal's immune status is shifted. For example, lysine represents a major component of body tissue proteins (65 to 70 g/kg) but a relatively small component of the proteins (24 g/kg) associated with maintenance functions (Fuller et al., 1989; Wang and Fuller, 1989). In contrast, sulphur amino acids (SAA) represent a small proportion of tissue proteins (16 g/kg) and a large proportion of proteins (49 g/kg) associated with maintenance functions. Thus, factors that increase tissue growth relative to maintenance needs in pigs would raise the animal's dietary lysine and methionine needs, but the magnitude of the change would be greater for lysine. The observation that the optimum ratio of digestible SAA to digestible lysine in pigs is less in animals experiencing a low versus high level of immune system activation supports the hypothesis (Table 10.4).

**Table 10.4** IMPACT OF LEVEL OF CHRONIC IMMUNE SYSTEM (IS) ACTIVATION ON DIETARY DIGESTIBLE LYSINE (L) AND SULPHUR AMINO ACID (SAA) NEEDS OF PIGS[a]

| Pig Weight, kg | IS Activation | Amino acid need, g/kg diet[b] | | Ideal SAA:L[c] |
|---|---|---|---|---|
| | | L | SAA | |
| 9 | Low | 13.4 | 6.4 | 0.48 |
| | High | 10.7 | 5.9 | 0.55 |
| 14 | Low | 12.2 | 6.2 | 0.51 |
| | High | 9.9 | 5.9 | 0.58 |

[a]Adapted from Williams and Stahly (1995).
[b]Digestible amino acid needs determined from break point analysis of gain:feed data.
[c]Ideal ratio of amino acids based on the dietary concentrations of digestible L and SAA needed to optimize gain:feed ratio.

## Impact of immune system activation on responses of pigs to specific feed ingredients

Certain feedstuffs possess properties that control or destroy antigens and/or result in the animal being less susceptible to an antigen challenge. Responses of pigs to these feedstuffs are dependent on the immune status of the animals being fed. For example, blood (plasma) collected from pigs and cattle and spray-dried contains relatively high concentrations of proteins, particularly immunoglobulins. Immunoglobulins are proteins that function to inactivate or destroy antigens in the body. Dietary additions of spray-dried plasma proteins result in improved daily gains and efficiency of feed utilization in antigen-challenged (high IS) pigs but not in animals with a low degree of antigen (low IS) exposure (Table 10.5).

Certain antimicrobial agents have been reported to lower intestinal cytokine release in antigen-challenged animals (Roura *et al.*, 1992). Based on the relationships of antigen exposure and body metabolism, antimicrobial agents that effectively control or destroy major antigens should enhance carcass leanness (muscle to fat ratio), as well as increase body growth and efficiency of feed utilization. Furthermore, the magnitude of the responses to these agents should be dependent on the animal's degree of antigen challenge. These results have been observed in growing pigs (Stahly *et al.*, 1994b; Stahly *et al.*, 1995). For example, pigs receiving subtherapeutic levels of an antimicrobial agent from 5 to 34 kilograms body weight grew faster, required less feed per unit of body weight and muscle gain, and produced carcasses with less fat and more dissectible muscle at market weight (Table 10.6). These responses occurred in pigs with a low and high level of chronic IS activation, but the magnitude of the response was greater in the pigs experiencing the greater antigen challenge.

**Table 10.5** RESPONSES OF PIGS WITH A LOW OR HIGH LEVEL OF ANTIGEN EXPOSURE TO DIETARY ADDITIONS OF PLASMA PROTEINS[a]

| Item | Antigen exposure | *Dried plasma proteins, g/kg* | | Unit change |
|---|---|---|---|---|
| | | *0* | *60* | |
| Daily feed, g | Low | 572 | 581 | +9 |
| | High | 449 | 536 | +87 |
| Daily gain, g | Low | 400 | 409 | +9 |
| | High | 245 | 327 | +82 |
| Feed/gain | Low | 1.44 | 1.42 | -.02 |
| | High | 1.88 | 1.66 | -.22 |

[a]Adapted from Stahly *et al.* (1994a). Pigs self-fed diets from 6 to 14 kilograms body weight.

**Table 10.6** RESPONSES OF PIGS WITH A LOW OR HIGH LEVEL OF ANTIGEN EXPOSURE TO DIETARY ANTIMICROBIAL AGENTS[a]

| Item | Antigen Exposure | *Carbadox, ppm* | | Unit Change |
|---|---|---|---|---|
| | | *0-0* | *55-0* | |
| Growth and feed utilization | | | | |
| Daily gain, g | Low | 845 | 858 | +13 |
| | High | 686 | 735 | +49 |
| Feed/gain | Low | 2.75 | 2.67 | -.08 |
| | High | 3.12 | 2.90 | -.22 |
| Carcass traits | | | | |
| Backfat, mm | Low | 27.4 | 26.4 | -1.0 |
| | High | 32.2 | 27.7 | -4.5 |
| L. muscle area, cm$^2$ | Low | 38.4 | 39.5 | +1.1 |
| | High | 32.6 | 35.8 | +3.2 |
| Carcass muscle, %[b] | Low | 56.0 | 57.6 | +1.6 |
| | High | 51.9 | 54.2 | +2.3 |

[a]Adapted from Stahly *et al.* (1994b). Pigs self-fed basal diet containing 0 or 55 ppm carbadox from 5 to 34 kilograms body weight and then all pigs placed on the control diet (0 ppm carbadox) from 34 to 115 kilograms body weight.
[b]Carcasses were physically dissected into muscle, bone, skin, fat tissues.

# Summary

Pigs possess an enormous potential to deposit muscle rapidly and efficiently when the animals' exposure to antigens is minimized. The optimum dietary regimen, both in terms of dietary nutrient concentrations and feed ingredient selection is influenced by the immune system status of the pigs being fed.

# Acknowledgements

The contributions of Noel Williams and Scott Swenson in conducting and analyzing portions of the experimentation presented in this manuscript are appreciated. Journal Paper No. J-16897 of the Iowa Agriculture and Home Economics Experiment Station, Ames, Iowa, Project No. 3142, and supported by Hatch Act and State of Iowa funds.

# References

Ballmer, P.E., McNurlan, M.A., Southorn, B.G., Grant, I. and Garlick, P.T. (1991). Effects of human recombinant interleukin-1 beta on protein synthesis in rat tissues compared with classic acute phase reaction induced by turpentine. *Biochem. J.* **279**:683.

Fan, J., Molina, P.E., Gelato, M.C. and Lang, C.H. (1994). Differential tissue regulation of insulin-like growth factor-I content and binding proteins after endotoxin administration. *Endocrinology.* **34**:1685.

Fuller, M.F., McWilliam, R., Wang, T.C. and Giles, L.R. (1989). The optimum dietary amino acid pattern for growing pigs. 2. Requirements for maintenance and tissue protein accretion. *Brit. J. Nutr.* **62**:255.

Hannager, S. Spagnoli, A., D'Urso, R., Navarra, P., Tsagarkis, S., Besser, M. and Grossman, A. B. (1991). Interleukin-1 modulates the acute release of growth hormone-releasing hormone and somatostatin from rat hypothalmus in vitro, whereas tumor necrosis factor and interleukin-6 have no effect. *Endocrinology.* **129**:1275.

Hashimoto, H., Igarashi, N., Yachie, A., Migawaki, T. and Sato, T. (1994). Relationship between serum levels of interleukin-6 and thyroid hormone in children with acute respiratory infection. *J. Clin. Endocrinol.* Metab. **78**:288.

Johnson, R.W. and von Borell, E. (1994). Lipopolysaccharide-induced sickness behavior in pigs is inhibited by pretreatment with indomethacin. *J. Anim. Sci.* **72**:309.

Klasing, K.C., Laurin, D.E., Peng, R.K. and Frey, D.M. (1987). Immunologically mediated growth depression in chicks: influence of feed intake, corticosterone and interleukin-1. *J. Nutr.* **117**:1629.

Mrosovsky, N., Molony, L.A., Conn, C.A. and Kluger, M.J. (1989). Anorexic effects of interleukin-1 in the rat. *Amer. J. Physiol.* **257**:R135.

Navarra, P., Tsagarakis, S., Fairia, M., Rees, L.H., Besser, G.M. and Grossman, A.B. (1991). Interleukins-1 and -6 stimulate the release of corticotropin-releasing hormone-41 from rat hypothalmus in vitro via the eicosanoid cyclooxygenase pathway. *Endocrinology.* **128**:32.

Roura, E., Homedes, J. and Klasing, K.C. (1992). Prevention of immunologic stress contributes to the growth-permitting ability of dietary antibiotics in chicks. *J. Nutr.* **122**:2383.

Shuster, D.E., Harmon, R.T., Jackson, J.A. and Hemkin, R.W. (1991a). Reduced lactational performance following intravenous endotoxin administration to dairy cows. *J. Dairy Sci.* **74**:3407.

Shuster, D.E., Harmon, R.T., Jackson, T.A. and Hemkin, R.W. (1991b). Suppression of milk production during endotoxin-induced mastitis. *J. Dairy Sci.* **74**:3763.

Stahly, T.S., Williams, N.H. and Zimmerman, D.R. (1995). Impact of tylosin on rate, efficiency and composition of growth in pigs with a low or high level of immune system activation. *J. Anim. Sci.* **73** (Suppl. 1):84.

Stahly, T.S., Swenson, S.G., Zimmerman, D.R. and Williams, N.H. (1994a). Impact of porcine plasma proteins on postweaning growth of pigs with a low and high level of antigen exposure. *ISU Swine Research Report, Iowa State University, Ames. AS-629*, pp 3-5.

Stahly, T.S., Williams, N.H. and Zimmerman, D.R. (1994b). Impact of carbadox on rate and efficiency of lean tissue accretion in pigs with a low or high immune system activation. *J. Anim. Sci.* **72** (Suppl. 1):165.

Wang, T.C. and Fuller, M.F. (1989). The optimum dietary amino acid pattern for growing pigs. 1. Experiments by amino acid deletion. *Brit. J. Nutr.* **62**:77.

Williams, N.H. and Stahly, T.S. (1995). Impact of immune system activation on the lysine and sulfur amino acid needs of pigs. *ISU Swine Research Report, Iowa State University, Ames, AS-633*, pp. 31-34.

Williams, N.H., Stahly, T.S. and Zimmerman, D.R. (1994). Impact of immune system activation on growth and amino acid needs of pigs from 6 to 114 kg body weight. *J. Anim. Sci.* **72** (Suppl. 2):57.

Williams, N.H., Stahly, T.S. and Zimmerman, D.R. (1993a). Impact of immune system activation and dietary amino acid regimen on nitrogen retention in pigs. *J. Anim. Sci.* **71** (Suppl. 1):171.

Williams, N.H., Stahly, T.S., Zimmerman, D.R. and Wannemuehler, M. (1993b). Impact of immune system activation on the amino acid needs of pigs. *J. Anim. Sci.* **71** (Suppl. 1):61.

Zamir, O., O'Brian, W., Thompson, R., Bloedow, D.C., Fischer, J.E. and Hasselgren, P. (1994). Reduced muscle protein breakdown in septic rats following treatment with interleukin-1 receptor antagonist. *Int. J. Biochem.* **26**:943.

**11**

# DIGESTIVE AND METABOLIC UTILIZATION OF DIETARY ENERGY IN PIG FEEDS: COMPARISON OF ENERGY SYSTEMS

J. NOBLET
*Institut National de la Recherche Agronomique, Station de Recherches Porcines, 35590 St Gilles, France*

## Introduction

The cost of feed is at least 50% of the total cost of pig meat production with the energy component representing the greatest proportion. Therefore, from a practical point of view, it is important to estimate precisely the energy value of feeds for least-cost formulation, for matching feed supply to energy requirements of animals or for legislation. However, a given diet or ingredient is ascribed different energy values according, initially, to the sequence of energy utilization by the pig (digestible (DE), metabolizable (ME) and net (NE) energy) and, secondly, to the prediction method used for each step. An energy system corresponds to the combination of one step in this sequence of energy utilization and one prediction method. The main difficulty is then to choose between available energy estimates for a given feed, its value compared to other feeds being dependent on the system employed. Thus, NE seems preferable since it takes into account the metabolic utilization of energy. In addition, NE is the only system in which energy requirements and diet energy values are expressed on the same basis which should be independent of the feed. Undoubtedly the adequacy of an energy system will be assessed through its ability to predict the animal performance with a satisfactory degree of accuracy and the energy value of both raw materials and compound feeds.

The objectives of this review are 1/ to discuss some methodological aspects, 2/ to consider the main factors of variation of digestive and metabolic utilization of energy in pig diets, 3/ to present the available energy systems for pig feeds with more emphasis on NE systems, 4/ to compare the energy systems and 5/ to evaluate some practical consequences for both diets formulators and pig producers. A complementary review was given by Noblet and Henry (1993). Information can also be obtained from several other reviews on this topic (Morgan, Whittemore, Phillips and Crooks, 1987; Henry, Vogt and Zoiopoulos, 1988; Batterham, 1990).

## Methodological aspects

Gross energy of feed (GE) is not totally available since some energy is lost in faeces, in urine, as gaseous products of fermentation (methane, hydrogen) and as heat (Figure 11.1).  Different ratios are calculated in order to appreciate the efficiencies at each level: digestibility coefficient (DC) of energy (DCe) is equivalent to the ratio between DE and GE; ME:DE is equal to the ratio between ME and DE and k is the efficiency of utilization of ME for NE.  Mean energy losses, expressed as percentages of GE and mean efficiencies are given in Figure 11.1.

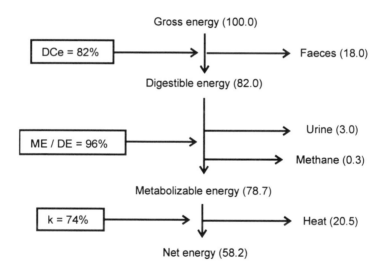

**Figure 11.1.** Energy utilization in pigs (values given are mean ratios obtained by Noblet *et al.*, 1994a on 61 diets)

The DE content of a feed corresponds to its GE content minus energy losses in faeces. However, DE is not a true measure of the feed energy absorbed from the digestive tract since faeces contain endogenous losses (digestive secretions, intestinal cell debris).  In addition, gas and heat from hindgut fermentation are produced but, when they are measured, they are considered in the further steps of energy utilization.  Nevertheless, quantification of endogenous and heat of fermentation losses is difficult and has a small practical interest. The DE concept used in practice corresponds therefore to the apparent DE content of a feed.

The ME content of a feed is equivalent to the difference between DE content and energy losses in urine and gases. Again, ME corresponds to an apparent value since a large proportion of the energy losses in urine originates from endogenous processes. Most of the energy lost in gases is due to methane production. While energy content of feed, faeces and urine can be measured with pigs kept in metabolism crates, the measurement of methane production necessitates the pig to be housed in a respiration chamber. Consequently, most ME values reported in the literature and tables ignore energy losses as methane.

NE is defined as ME content minus heat increment (HI) associated with metabolic utilization of ME and the energy cost of ingestion and digestion of the feed. In growing or fattening pigs fed above their maintenance energy requirements, only a fraction (so-called $k_g$) of the additional ME supply ($\Delta ME$), is retained in the body and the other portion which corresponds to HI, is lost as heat. Similarly, below maintenance energy supply, $\Delta ME$ is used for sparing energy from body reserves with an associated HI ($HI_m$); the efficiency of utilization of ME ($k_m$) corresponds then to the ratio ($\Delta ME$ - $HI_m$)/$\Delta ME$ (Figure 11.2). But, in practice, only the amount of total energy retained (RE) or total heat production (HP) can be measured directly in animals and $k_g$ (or $k_m$) or HI (or $HI_m$) correspond to the calculated slopes of the relationships between RE or HP and ME intake (Figure 11.2).

As discussed below, HI of a given feed is not constant over a large range of ME intakes and it depends on physiological factors. For instance, it tends to be lower below than above maintenance energy supply; it is also lower when ME is used for fat deposition than for protein gain. Therefore, for comparing different feeds for their HI or their efficiencies of utilization of ME ($k_g$ or $k_m$), it is essential to calculate these values under comparable conditions such as similar feeding level and constant composition of the gain.

Logically, it would be necessary to feed the pigs different energy levels in order to calculate the values of HI or k of a given feed. Such measurements are complex and time-consuming, so that only one feeding level is usually applied. The slopes are then calculated by taking an estimate of fasting heat production (FHP) (Figure 11.2). The value of FHP is either calculated by extrapolating HP as measured at different energy levels to zero ME intake or estimated from literature. For animals in positive energy balance animals, the NE value corresponding to total ME intake from a feed is then equivalent to the sum of estimated FHP and RE. Such an approach means that, in such animals, the efficiencies of ME for gain and for maintenance are assumed to be the same for a given diet. It also means that the NE value of a feed is directly dependent on the estimate of FHP or, in other words, on the factors responsible for variation of maintenance energy requirements (environment, physical activity).

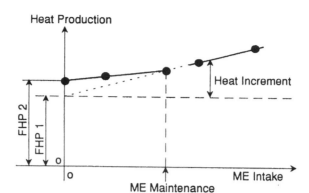

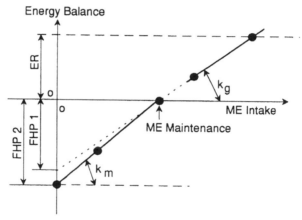

**Figure 11.2.** Methodology for estimating NE value in pig feeds; FHP1 corresponds to fasting heat production in positive energy balance pigs calculated by extrapolation to zero ME intake and FHP2 to measured (or estimated by extrapolation) fasting heat production in below maintenance fed pigs; ER: Energy retained

From a practical point of view and in order to avoid any bias in the calculation of NE of a series of feeds, it is then strongly recommended to carry out energy balance measurements in pigs from the same sex, the same breed and the same body-weight, to keep them in a temperature-controlled environment (above their critical temperature) and to feed them at the same energy level. Under such circumstances, an erroneous estimate of FHP will affect the absolute NE value but not the hierarchy between feeds within a series. Energy balance can be measured either according to the comparative slaughter technique (direct measurement of retained energy) or the calorimetry methods (measurement of heat production). The most commonly used method is indirect calorimetry which consists of the calculation of heat production from gas exchanges.

While measurements of DE value and, to a smaller extent, of ME value are easy and can be undertaken on a large number of feeds at a reasonable cost, measurement of NE is more complex and expensive. The principle is then to carry out a series of NE measurements on different diets and to combine all results in prediction equations of feed NE content, the predictors to be considered being available from digestibility experiments. Consequently, from a statistical point of view, it is important to design the NE experimental diets in order to have dietary chemical characteristics as independent as possible. Different prediction models can be used but they must be realistic. This means that predictors must be easily available, either from feding tables or at the laboratory level.

It is important to emphasise that each NE system corresponds to one NE prediction equation. This means that a given feed will have as many NE values as equations or NE systems and that two feeds, when compared on a NE basis, should be evaluated from the same equation. This also means that one equation established under given conditions (type of pigs, feeding level) could not theoretically be applied to other conditions.

Most aspects presented in this review will be based on a series of experiments undertaken over the last 10 years, with the final objective of proposing a NE system which could be applied at all stages of pig production and all types of pigs. An attempt was made to overcome most of the methodological problems which have been listed above. Three main trials were conducted. The first one (Noblet, Fortune, Shi and Dubois, 1994a) considered 45 to 50 kg fast-growing and lean pigs (Large White boars) as a model of modern pigs; they were fed the same energy level (close to *ad libitum*: 2.3 MJ ME/kg$^{0.60}$) and all kept at 22°C. Their FHP was calculated as 750 kJ/kg$^{0.60}$ A total of 61 diets which differed widely in their chemical characteristics were measured. In a second trial, 14 diets (from the 61 diets given to growing pigs) were fed to adult sows at and below their maintenance energy requirements in order to study the effect of diet composition on HIm or $k_m$ and propose prediction equations of NE for maintenance (Noblet, Shi and Dubois, 1993). These values could be compared to corresponding values obtained in growing pigs. In a third trial, the effect of body weight (or of composition of body weight gain) was studied by feeding 7 diets to 45-50 kg, 100 kg and 150 kg pigs (Noblet, Shi and Dubois, 1994b).

## Energy utilization

DIGESTIVE UTILIZATION

For most pig diets, DCe varies between 70 and 90%; for raw materials, variations are larger (0 to 100%). These variations are associated with differences in faecal digestibility of the nutrients constituting organic matter. With regard to crude protein and crude fat, their (true) DC vary between 60 and 95% according to their chemical characteristics and their origin, while soluble carbohydrates (starch and sugars) are highly digestible (95 to 100%). In fact, most of the variation of DCe is associated with the presence of fibre (defined as the sum of non-starch polysaccharides (NSP) and lignin) which is less digestible (below 50%) and also reduces the apparent faecal digestibility of crude protein and fat (Noblet and Shi, 1993; Noblet and Perez, 1993). In addition, the digestive utilization of fibre is variable: for instance, Chabeauti, Noblet and Carré (1991) found DCe of total NSP in wheat straw, wheat bran, sugar beet pulp and soybean hulls equivalent to 16, 46, 69 and 79%, respectively. To a lesser extent, the amount of minerals has also a negative effect on DCe (Table 11.1).

According to these observations, the reduction of DCe with dietary fibre addition will vary with the fibrous material evaluated (Perez, Ramihone and Henry, 1984; Chabeauti *et al.*, 1991; Figure 11.3). Therefore, the amount of total dietary fibre is an inadequate criterion for predicting DCe; additional information on chemical and physical characteristics would then be necessary. Description and limits of chemical methods for measuring dietary fibre have been discussed by Noblet and Henry (1993). Furthermore, the coefficient attributed to fibre in Table 11.1 indicates that, at least in growing pigs, NDF is equivalent to a diluent (1% decrease of DCe per 1% NDF) while NDF is 40 to 50% degraded in the digestive tract. In other words, even dietary fibre can be degraded by the pig, it does not provide a significant amount of available energy mainly because degradation of fibre is associated with excretion of endogenous protein and fat (Noblet and Shi, 1993).

**Table 11.1** EFFECT OF DIET COMPOSITION (G/KG DRY MATTER) ON DIGESTIBILITY COEFFICIENT OF ENERGY (DCe, %) AND ME:DE RATIO IN GROWING PIGS OR ADULT SOWS[1].

| No | Equation | RSD | Reference[2] |
|----|----------|-----|-----------|
| 1 | DCe = 102.1 - 0.171 x Ash - 0.167 x CF | 2.2 | 1 |
| 2 | DCe = 101.3 - 0.095 x Ash - 0.095 x NDF | 1.7 | 1 |
| 3 | DCe = 100.5 - 0.079 x Ash - 0.088 x NDF - 0.118 x ADL | 1.5 | 1 |
| 4 | DCe = 99.3 - 0.076 x NDF | 1.6 | 2 |
| 5 | DCe = 101.4 - 0.133 x NDF | 1.4 | 2 |
| 6[3] | ME:DE = 100.7 - 0.021 x CP - 0.005 x NDF | 0.5 | 3 |
| 7 | ME:DE = 100.3 - 0.021 x CP | 0.5 | 1 |

[1] CF: Crude Fibre, CP: crude protein, NDF: Neutral Detergent Fibre, ADL: Acid Detergent Lignin.

[2] 1: Noblet and Perez, 1993 (45 kg pigs; n=114 diets); 2: Noblet and Shi, 1993 (equation 4: adult sows; equation 5: 45 kg pigs; n=14 diets); 3: Noblet *et al.*, 1989a (45kg pigs; n = 41 diets).

[3] Methane energy losses included

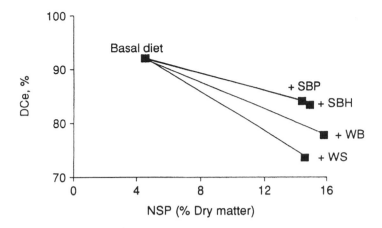

**Figure 11.3**. Effect of sugarbeet pulp (SBP), soybean hulls (SBH), wheat bran (WB) and wheat straw (WS) addition to a basal diet on digestibility coefficient of energy (DCe) in growing pigs (from Chabeauti *et al.*, 1991)

Limits of analytical procedures and variations in digestive utilization of dietary fibre explain the unsatisfactory and contradictory relationships between DCe and estimate of fibre content. However, for practical purposes, CF or NDF and/or ADF represent reasonable predictors of DCe in diets in which several fibrous sources are mixed (Table 11.1), with a preference for NDF (Morgan *et al.*, 1987; Noblet, Fortune, Dubois and Henry, 1989; Noblet and Perez, 1993). Such equations should not be applied to raw materials where specific relationships are to be used (Noblet and Henry, 1993).

Apart from the effect of antinutritional factors and heat treatments that will not be considered in the present paper, DCe is affected by factors other than those related to the diet itself. Initially, in growing pigs, DCe decreases when feeding level is elevated (Everts, Smits and Jongbloed, 1986; Roth and Kirchgessner, 1984; Noblet, Shi, Karege and Dubois, 1993a). Such an effect, which is small from a practical point of view, is not observed in adult sows at feeding levels ranging from 2 to 3.6 kg per day (J. Noblet, unpublished data). In growing pigs, DCe increases with body weight (BW), with larger differences with high fibre diets or raw materials (Figure 11.4). Similar interactions between BW of pigs and diet characteristics were observed over a wider BW range by Noblet and Shi (1994). From a literature review, it was calculated that, over the 30 to 100 kg BW period, DCe is increased by 0.30 and 0.45 unit for each 10 kg increase in BW for diets containing 4 and 6% crude fibre, respectiveley (Noblet *et al.*, 1993a). The largest effect of BW is observed when adult sows and growing pigs are compared: digestibility coefficients are superior in all cases for the sows, the difference being greater with fibrous diets or ingredients (Figure 11.5 and Table 11.2). In addition, the DE difference between sows and growing pigs for a given level of dietary fibre, depends on the origin of fibre (comparison of corn gluten feed and wheat bran in Table 11.2) with subsequent variations in the DE difference of raw materials between sows and growing pigs. At similar feeding levels, DCe is not significantly affected by pig genotype or climatic environment (Noblet, Le Dividich and Bikawa, 1985; Noblet *et al.*, 1993a).

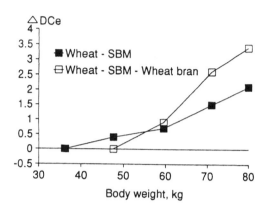

**Figure 11.4**. Interaction between body weight of pigs and diet characteristics (wheat + soybean meal vs wheat + soybean meal + wheat bran) on variation of digestibility coefficient of energy over the growing period (ΔDCe: difference between DCe and DCe measured during and at the beginning of the experiment respectively; J. Noblet, unpublished data)

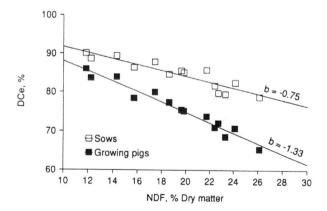

**Figure 11.5**. Effect of diet NDF content on digestibility coefficient of energy (DCe) in sows fed at maintenance level and 45 kg growing pigs fed ad libitum (from Noblet and Shi, 1993)

One practical consequence of these changes of DCe with BW is that digestibility trials representative of the 'growing-finishing' period should be carried out at about 60 kg BW. Another consequence is that at least two different DE values should be given to feeds, one for growing pigs and one for adult sows (Noblet *et al.*, 1993a).

**Table 11.2** COMPARATIVE DE VALUES OF DIETS OR INGREDIENTS IN ADULT SOWS AND GROWING PIGS

|  | DE in sows, MJ/kg DM | DE in growing pigs % of DE in sows |
| --- | --- | --- |
| *Experiment 1*[1] |  |  |
| Diets (n=14) | 15.6 | 89.4 (83.1-95.5)[3] |
| Wheat | 17.0 | 100 |
| Wheat bran | 13.1 | 81 |
| Corn gluten feed | 17.4 | 72 |
| Sugarbeet pulp | 11.4 | 61 |
| *Experiment 2*[2] |  |  |
| Diets (n=13) | 15.4 | 96.3 (94.0-98.3)[3] |
| Wheat + soyabean meal | 16.6 | 98.3 |
| Wheat bran | 12.4 | 90.5 |
| Corn gluten feed | 13.8 | 72.2 |

[1] Adult sows fed 1.5 kg dry matter per day *vs* 45 kg growing pigs fed 1.4 kg dry matter per day (after Noblet and Shi, 1993 and Shi and Noblet, 1993)
[2] Adult sows fed 2.1 kg dry matter per day *vs* 65 kg growing pigs fed 2.0 kg dry matter per day (J. Noblet, unpublished data)
[3] Range for all diets

ME:DE RATIO

In growing pigs, average energy loss as methane is equivalent to 0.4% of DE intake (Noblet *et al.*, 1994a), with values ranging from 0.1 to 1.2%; the latter value is obtained with diets which contained soybean hulls or sugar beet pulp (highly digestible fibre). In that particular situation, the methane energy loss represented about 5% of the DE of the raw material. In sows fed at maintenance level, energy of methane averages 1.5% of DE intake (Noblet and Shi, 1993).

Energy lost in urine represents a variable percentage of DE since urinary energy is highly dependent on the amount of nitrogen in urine (3 kJ per g of N x 6.25; Noblet and Perez, 1993). At a given physiological stage where the amount of nitrogen retained in the body is stable, the urinary nitrogen will depend mainly on the amount of digestible protein and, therefore, on the crude protein (CP) content of the diet. Consequently, the ME:DE ratio is linearly related to dietary protein content (Table 11.1). Between physiological stages, protein retention as a percentage of digestible or dietary protein may vary to a large extent. For instance, in adult sows fed at maintenance and 45 kg growing pigs fed *ad libitum*, urinary energy represented 6.5 and 3.3% of DE, respectively, when both groups were given the same diets (Noblet and Shi, 1993b).

In most situations, the ME:DE ratio is relatively constant and equivalent to about 0.96. However, that ratio is not acceptable when dietary CP content and/or protein retention are either high or low. In addition, that mean value cannot be applied to single feed ingredients. For instance, ME:DE ratios (methane energy loss included) ranged from 100% for animal fat to 97-98% for cereals, 93 to 96% for protein sources (soybean meal, peas) and 90 to 92% for fibrous protein sources (rapeseed meal, sunflower meal) (Noblet, Fortune, Dupire and Dubois, 1993b). Equations presented in Table 11.1 should then be used.


## METABOLIC UTILIZATION OF ME

NE is defined as ME minus heat increment associated with metabolic utilization of ME and also to the energy cost of ingestion and digestion. The ratio between NE and ME (or k) varies initially according to the final utilization of energy. With standard cereals-soybean meal diets, it was 72% for milk energy in sows ($k_l$) (Noblet, Dourmad and Etienne, 1990), 75% for energy storage in growing pigs ($k_g$), 80% and 60% for energy deposition as fat ($k_f$) and as protein ($k_p$), respectively (Noblet, Karege and Dubois, 1991) and 77% for meeting maintenance requirements (Noblet *et al.*, 1993c). Secondly, the efficiency of ME for NE varies according to the chemical characteristics of the feed since nutrients (carbohydrates, amino acids, long-chain fatty acids or volatile fatty acids) are not used with similar efficiencies (Table 11.3).

**Table 11.3** EFFECT OF DIET COMPOSITION (G/KG DRY MATTER) ON EFFICIENCY OF UTILIZATION OF METABOLIZABLE ENERGY FOR NET ENERGY IN GROWING PIGS ($K_G$) AND MAINTENANCE FED ADULT SOWS ($K_M$)[1].

| No | Equation | $R^2$ | Reference |
|----|----------|-------|-----------|
| 5 | $k_g$ = 74.7 + 0.036 x EE + 0.009 x ST - 0.023 x CP - 0.026 x ADF | 0.66 | Noblet *et al.*, 1994a |
| 6 | $k_m$ = 67.2 + 0.066 x EE + 0.016 x ST | 0.57 | Noblet *et al.*, 1994b |

[1] See Table 11.1 for abbreviations; EE: ether extract, ST: starch, ADF: Acid Detergent Fibre.

The variations of $k_g$ and $k_m$ with dietary chemical characteristics are due to differences in efficiencies of ME utilization between nutrients, with a similar hierarchy between nutrients for maintenance and for growth (Table 11.4). In near mature or mature pigs depositing predominantly fat, Schiemann, Nehring, Hoffmann, Jentsch and Chudy (1972) obtained efficiencies (i.e. for maintenance + fattening) of 52, 63, 98 and 73% when ME was provided by digestible protein (DCP), digestible crude fibre (DCF), digestible fat (DFat) and digestible nitrogen-free extract (NFE equivalent to organic matter-(protein+crude fibre+fat)), respectively. Similarly, in growing pigs (i.e. for maintenance + growth), Just and co-workers showed that the efficiency of utilization of ME increased when more fat was included in the diet (Just, 1982a) and decreased with increased protein (Just, 1982b) or crude fibre (Just, 1982c). Even if the results of these studies are not directly comparable, they all indicate a low efficiency of utilization of end-products of fibre digestion (volatile fatty acids) or energy digested at the hindgut level (Table 11.4). In spite of differences in the composition of energy gain with body weight and higher values for $k_f$ than for $k_p$, no significant effect of BW (in the 45 to 150 kg BW range) on $k_g$ was observed (Noblet *et al.*, 1994b). Finally, $k_g$ is dependent on climatic conditions with higher values in the cold and a higher change with temperature variation for low efficiency diets (fibre- or protein-rich) (Noblet *et al.*, 1985; Noblet, Dourmad, Le Dividich and Dubois, 1989b).

**Table 11.4**  EFFICIENCIES (%) OF UTILIZATION OF DIGESTIBLE (DE) OR METABOLIZABLE (ME) ENERGY OF NUTRIENTS FOR NET ENERGY (NE) FOR GROWTH OR MAINTENANCE[1].

| | NE/ME, % | | NE/DE, % | |
|---|---|---|---|---|
| | *Growth* | *Maintenance* | *Growth* | *Maintenance* |
| Diets | 73.9 | 77.4 | 71.0 | 71.4 |
| Digestible nutrients: | | | | |
| Crude protein | 58 | 69 | 50 | 60 |
| Ether extract | 90 | 102 | 90 | 100 |
| Starch | 82 | 82 | 82 | 81 |
| Sugars | 73 | 82 | 72 | 81 |
| NDF - ADF | 66 ⎫ | - ⎫ | 69 ⎫ | - ⎫ |
| ADF | 0 ⎬ 58[3] | - ⎬ 56[3] | 0 ⎬ 54[3] | - ⎬ 43[3] |
| Residue[2] | 77 ⎭ | - ⎭ | 70 ⎭ | - ⎭ |
| Illeal DE | 76 | 82 | 74 | 80 |
| Hindgut DE | 58 | 59 | 54 | 46 |

[1]  After Noblet *et al.* (1994a and 1993c)
[2]  Digestible residue corresponds to the difference between digestible organic matter and the sum of the other digestible nutrients.
[3]  Mean value for total fibre (Noblet *et al.*, 1994a and 1993c).

**Energy systems**

DIGESTIBLE ENERGY

The DE content of a diet can be obtained directly with pigs kept in metabolism cages from determination of dietary and faecal energy. This method has been widely used for measurement of DE of raw materials reported in feeding tables. However, that approach cannot be used in routine measurements on a large number of samples and where a rapid response is required. Apart from taking mean values given in tables for raw materials, alternative methods have therefore been proposed.

For raw materials, a first approach is to relate the DE of a feed to its content of digestible nutrients. This method involves measurement of gross chemical composition and estimation of digestibility coefficients of nutrients, the DE content being predicted from regression equations (Table 11.5). Tables giving DC of nutrients according to the Weende procedure (CP, CF, Fat and NFE) are available (DLG, 1984; CVB, 1988). In comparison with mean tabulated values for raw materials, this method takes into account the variations in chemical composition of some ingredients. A second approach is to use specific equations for prediction of DE value of ingredients of highly variable composition (Noblet and Henry, 1993). Some published equations are presented in Table 11.5. However, no equation is available for many ingredients.

The DE content of compound feeds can be obtained by adding the DE contributed by individual ingredients and assuming no interaction. When the actual composition of the feed is unknown, the only possibility is to use prediction equations based on chemical criteria. Some of the numerous proposed equations are presented in Table 11.5. In all equations, fibre has an important effect on the accuracy of the prediction. But as pointed out in the above section, the main limitation of such equations is their inability to consider the nature of fibre and, to a smaller extent, the composition of fat. The main consequence is that, according to the fibre source, the DE value of the diet when predicted from such equations will be either overestimated when fibre is poorly digestible (from wheat straw, for instance) or underestimated when fibre is highly digestible (from sugar beet pulp or soybean hulls, for instance). In all equations, ash content has a significant and negative contribution to DE. However, the coefficient assigned to ash is usually higher than what would be expected from the dilution effect of ash (Just, Jorgensen and Fernandez, 1984; Perez *et al.*, 1984; Morgan *et al.*, 1987; Noblet and Perez, 1993) (Table 11.4). No clear physiological explanation for this result can be given.

As illustrated in the above section, DCe is affected by feeding level and body weight and physiological stage of the animals. It would therefore be logical to use DE values adapted to each situation. From a practical point of view, use of two DE values is suggested for most pig raw materials, one for '60 kg' pigs and one

**Table 11.5** PREDICTION EQUATIONS OF DE OR ME CONTENTS (MJ/KG DRY MATTER) OF PIG FEEDS (COMPOSITION: G/KG DRY MATTER)[1].

| No | Feed | Equation | RSD | $R^2$ | Reference |
|---|---|---|---|---|---|
| 9 | Ingredients + diets | DE = 0.0229 x DCP + 0.0389 x DEE + 0.0115 x DCF + 0.0175 x ST + 0.0169 x Sugars + 0.0183 x DRes | 0.09 | 0.99 | J. Noblet unpublished |
| 10 | Ingredients + diets | ME = 0.0200 x DCP + 0.0394 x DEE + 0.0096 x DCF + 0.0175 x ST + 0.0167 x Sugars + 0.0173 x DRes | 0.11 | 0.99 | J. Noblet, unpublished |
| 11 | Barley | DE = 17.04 - 0.046 x CF | - | 0.92 | Perez et al., 1980 |
| 12 | Rapeseed meal | DE = 17.28 + 0.020 x EE - 0.030 x CF | - | 0.77 | Bourdon, 1986 |
| 13 | Sunflower meal | DE = 16.85 - 0.033 x CF | - | 0.92 | Perez et al., 1986 |
| 14 | Mixed diets | DE = 17.37 - 0.051 x Ash + 0.010 x CP + 0.016 x EE - 0.027 x CF | 0.35 | 0.89 | Noblet and Perez, 1993 |
| 15 | Mixed diets | DE = 17.44 - 0.038 x Ash + 0.008 x CP + 0.016 x EE - 0.015 x NDF | 0.28 | 0.92 | Noblet and Perez, 1993 |

[1] See Tables 11.1 and 11.2 for abbreviations; DCP: digestible CP, DEE: digestible ether extract, DCF: digestible crude fibre, ST: starch, DRes: digestible residue calculated as digestible organic matter minus other digestible nutrients considered in the equation.

for adult sows. Values given in feeding tables appear to be more adapted to the '60 kg pig'. Equations for predicting DE in sows from DE in growing pigs can then be used to estimate DE in sows. Such preliminary equations were proposed by Noblet and Shi (1993). Final ones which are adapted to groups of ingredients will be published in the near future.

## METABOLIZABLE ENERGY

The approaches for predicting ME value of pig feeds are similar to those described for DE. But since direct ME measurements are not carried out routinely, tabulated values have been usually calculated from DE values with a ME:DE ratio either constant or, preferably, related to the protein content of the diet (INRA, 1984; NRC, 1988). ME is also predicted from equations relating ME to digestible nutrients content (Just, 1982d; Table 11.5). Like DE, ME content of mixed diets can be estimated from chemical composition (Noblet and Perez, 1993). Finally, ME values can be corrected for a zero or constant N balance (de Goey and Ewan, 1975) or for the amount of fermented carbohydrates (DLG, 1984).

## NET ENERGY

All published NE systems combine the utilization of ME for maintenance and for growth (Just, 1982d: NEj; Noblet *et al.*, 1994a and 1994b: NEg) or for fattening (Schiemann *et al.*, 1972: NEs) by assuming similar efficiencies for maintenance and energy retention. The most important equations (or systems) are given in Table 11.6. Equations proposed by Noblet *et al.* (1993c) can be used to predict NE value of feeds in a maintenance situation. Apart from these equations obtained from measurements conducted on a large number of diets, NE of individual ingredients have been proposed by Ewan and co-workers from studies conducted in young piglets (de Goey and Ewan, 1975). An attempt to combine all data in a general equation has been made by Ewan (1989). A more complete description of these NE systems has been given by Noblet and Henry (1991 and 1993). The NEj system has been mainly used in Denmark and, to a limited extent, in some Northern Europe countries. The expansion of the NEs system is directly connected to its application in the Netherlands (CVB feeding tables). Recently, CVB proposed an equation which is a compromise between the NEs equation and a preliminary equation obtained for growing pigs (Noblet *et al.*, 1989a).

As indicated in Table 11.6, three different approaches were used for predicting NE from measurements conducted in 45-50 kg pigs. The first one was comparable to the technique used by Schiemann *et al.* (1972) but different fractionation methods

**Table 11.6**  EQUATIONS FOR PREDICTION OF NET ENERGY IN PIG FEEDS (MJ/kg DRY MATTER) (COMPOSITION AS g PER kg OF DRY MATTER)[1]

| No | Equation | RSD % | Reference |
|----|----------|-------|-----------|
| 16 | NEs = 0.0109 x DCP + 0.0361 x DEE + 0.0090 x DCF + 0.0125 x DRes | 3.8 | Schiemann *et al.*, 1972 |
| 17 | NEj = 0.75 x ME - 1.88 | 2.3 | Just, 1982d |
| 18 | NEnl = 0.0108 x DCP + 0.0361 x DEE + 0.0135 x ST + 0.0127 x Sugars + 0.0095 x DRes | - | CVB, 1993 |
| 19 | NEg2 = 0.0113 x DCP + 0.0350 x DEE + 0.0144 x ST + 0.0000 x DCF + 0.0121 x DRes | 2.0 | Noblet *et al.*, 1994a |
| 20 | NEg4 = 0.703 x DE - 0.0041 x CP + 0.0066 x EE - 0.0041 x CF + 0.0020 x ST | 1.7 | Noblet *et al.*, 1994a |
| 21 | NEg7 = 0.730 x ME - 0.0028 x CP + 0.0055 x EE - 0.0041 x CF + 0.0015 x ST | 1.6 | Noblet *et al.*, 1994a |
| 22 | NEg11 = 12.03 - 0.0230 x Ash + 0.0183 x EE - 0.0084 x (NDF-ADF) - 0.0168 x ADF + 0.0028 x ST | 2.6 | Noblet *et al.*, 1994a |

[1] See Tables 11.1, 11.3 and 11.5 for abbreviations; NEs: Schiemann et al. (1972), NEj: Just (1982d), NEnl: CVB, 1983, DRes corresponds to the difference between digestible organic matter and other digestible nutrients considered in the equation

of digestible nutrients were compared. The second one combines DE (or ME) content and some chemical characteristics as predictors. The objective of the third one was to predict NE of a diet from its chemical characteristics. The equations resulting from the first two approaches can be used for raw materials and diets; required information (DE or ME plus chemical characteristics or digestible nutrients) is available in most feeding tables. The equations based on crude nutrients are applicable only to complete diets; their accuracy is limited by the accuracy of dietary chemical analysis and origin of nutrients since, in such models, the variability in digestibility of some chemical fractions (fat and fibre) is ignored. Compared to the previous studies of Schiemann *et al.* (1972), the data demonstrate the advantage of a more appropriate fractionation of carbohydrates and particularly the importance of starch for prediction of NE. Studies also show that the prediction based on only ME concentration (Just, 1982d) can be improved by including additional chemical characteristics.

The studies conducted in heavier pigs (Noblet *et al.*, 1994b) indicate that the equations obtained in young growing pigs (NEg2, NEg4 or NEg7; Table 11.6) are applicable at other stages of the growing period. Indeed, covariance analysis showed no significant effect of stage of growth on NE and no interaction between the coefficients of each equation and the stage of growth (Noblet, Shi, Fortune, Dubois, Lechevestrier, Corniaux, Sauvant and Henry, 1994). In agreement with the superiority of $k_m$ over $k_g$ for the same set of diets (Noblet *et al.*, 1993c), the equations obtained for growing pigs (Table 11.6) give calculated NE values (or NEcal according to digestibility results obtained with sows) lower than measured NEm in adult sows. However, the difference between both values was not explained by any dietary chemical characteristic and the hierarchy between diets was the same for NEm and NEcal. In other words, the equations calculated from measurements conducted in growing pigs can be used for predicting NE of diets given to maintenance fed pigs. The subsequent underestimation of NEm can be taken into account in the expression of maintenance energy requirements. However, it must be repeated that different NE values should be used for 60 kg pigs and adult sows in connection with the higher DE content for sows.

## Comparison of energy systems

### DE, ME AND NE SYSTEMS

From equations reported in Table 11.6, it is obvious that the hierarchy between feeds obtained in the DE or ME systems will vary in the NE system according to their specific chemical composition. Since NE represents the best estimate of the "true" energy value of a feed, this means that the energy value of protein or fibrous feeds is overestimated when expressed on a DE (or ME) basis. On the other hand, fat or starch sources are underestimated in a DE system. These conclusions are clearly demonstrated in Tables 11.7 and 11.8 in the case of ingredients.

**Table 11.7** MEASURED AND CALCULATED ENERGY VALUES OF SOME INGREDIENTS (AFTER NOBLET *et al.*, 1993b AND 1994a)[1]

|  | *DE* | *ME* | *NE* | *NEs* | *NEj* | *NEnl* | *NEg* |
|---|---|---|---|---|---|---|---|
| 'Diet', MJ/kg DM[2] | 16.69 | 16.19 | 12.13 | 11.17 | 10.29 | 11.60 | 12.23 |
| As a percentage of 'diet' energy content | | | | | | | |
| Wheat | 97 | 97 | 100 | 97 | 97 | 99 | 99 |
| Corn | 95 | 94 | 102 | 99 | 93 | 101 | 101 |
| Tapioca | 95 | 96 | 106 | 102 | 95 | 105 | 104 |
| Soybean meal | 98 | 94 | 66 | 79 | 93 | 74 | 75 |
| Peas | 97 | 97 | 91 | 94 | 96 | 94 | 92 |
| Animal fat | 179 | 182 | 242 | 246 | 198 | 236 | 222 |

[1]  DE, ME and NE correspond to measured energy values and NEs, NEj and NEnl to calculated NE values (see Table 11.6); NEg is the mean of calculated NE values from equations NEg2, NEg4 and NEg7 (Table 11.6).
[2]  Diet corresponds to the combination of 81.5% wheat, 15% soybean meal and 3.5% animal fat.

**Table 11.8**  RELATIVE ENERGY VALUES OF SOME FEEDS IN THE DIGESTIBLE (DE) AND NET ENERGY (NE) SYSTEMS (AFTER NOBLET *et al.*, 1994b)

| System | *DE* | *NE* | *NE/DE (x 100)* |
|---|---|---|---|
| Basal diet, MJ/kg DM[1] | 15.62 | 11.08 | 71 |
| Ingredients, % of basal diet | | | |
| Corn starch | 115 | 131 | 82 |
| Rapeseed oil | 237 | 300 | 90 |
| Sucrose | 104 | 114 | 78 |
| Protein mixture[2] | 147 | 133 | 64 |
| Fibre mixture[3] | 55 | 44 | 57 |

[1]  Cereals, soybean meal and MV mixture
[2]  50% casein and 50% extracted soybean protein
[3]  25% wheat bran, 25% soybean hulls, 25% sugar beet pulps and 25% ground wheat straw

NET ENERGY SYSTEMS

As described above (Table 11.6), several equations (and therefore systems) for prediction of NE of feeds are available. In order to evaluate these systems, the NE values measured on 61 diets by Noblet *et al*., (1994a) have been compared with the calculated NE values from Schiemann *et al*. (1972) (NEs), Just (1982d) (NEj) and the equation proposed by CVB (1993) (NEnl) in the Netherlands. Values were also compared with those obtained by the equations proposed by Noblet *et al*. (1994a). The DE, ME and digestible nutrient contents of these diets were measured. A similar approach was conducted on data obtained by Noblet *et al*. (1993b) on a few feedstuffs (Table 11.7).

The mean calculated NE values for the 61 diets were 9.89, 8.76 and 10.12 MJ per kg dry matter in the NEs, NEj and NEnl systems, respectively, while the measured mean value was 10.50 MJ. The mean difference between measured and calculated NE values of diets is mainly due to differences in the estimate of FHP. However, the difference is also related to dietary chemical characteristics and not proportional to the energy content of the diet (Noblet *et al*., 1994a). The analysis of the differences indicates that NEs would underestimate the NE content of starch-rich feeds (Figure 11.6). When moving from NEs to NEnl, the underestimation of starch-rich feeds is reduced but it remains significant. With regard to the NEj system, it underestimates the NE value of starch- or fat-rich feeds and overestimates the NE value of fibre-or protein-rich feeds. In fact, the hierarchy between ingredients in the NEj system is closer to what is observed in the DE system than in the other NE systems (Table 11.7). As expected, no difference between NE calculated from equations published by Noblet *et al*. (1994a) and measured NE was obtained.

VALIDATION OF ENERGY SYSTEMS

Net energy equations are obtained from energy balance measurements conducted under specific conditions (animals, body gain composition, environment, experimental design, analytical procedures, statistical models). Each system should then be tested under other experimental conditions and also in practical conditions in order to validate the proposed equations.

It is obvious that the NE measurements carried out by Just and co-workers on growing pigs were correlated with growth performance since the comparative slaughter technique was applied over the total growth phase. The equations these authors proposed are therefore adapted to prediction of NE value of diets. However, as illustrated in the previous section, the equation is unable to predict accurately the NE of ingredients (Table 11.7) or extreme diets. A validation experiment of the NEs system, conducted on growing pigs (Borggreve, Van kempen, Cornelissen and Grimbergen, 1975), indicated that the dietary energy value was underestimated for high starch diets. This conclusion agrees with those obtained in the comparison of NEs and measured NE on diets or raw materials (Noblet *et al.*, 1994a; Figure 11.6). This problem is probably reduced in the NEnl system.

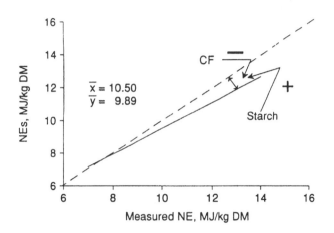

**Figure 11.6**. Relationship between net energy content calculated from Schiemann *et al.* (1972) (NEs) and measured net energy (measured NE) of 61 diets (from Noblet *et al.*, 1994a)

With regard to the NE equations proposed by Noblet *et al.* (1994a), an initial validation was given by the comparison of measured (in respiration chambers) NE contents of a few feedstuffs and their calculated NE values; they were quite comparable (Table 11.7). A similar approach applied to rather pure ingredients confirmed the hierarchy between nutrients for their DE or NE values (Table 11.8). A second validation was provided by results of growth trials. In a first experiment, growth performance (growth rate and body composition at slaughter) were measured in 540 pigs fed, according to an energy feeding scale, in order to obtain similar growth performance between diets. Eight diets were tested. The variability of the feed energy:BW gain ratio between diets (adjusted for similar BW gain, amount of gut fill and composition of the gain) was used as a criteria for evaluating each energy system (Noblet *et al.*, 1994c). The results confirm the superiority of NE equations published by Noblet *et al.* (1994a) over DE system and the bias due to starch level in the NEs system. Results of other recent experiments confirm the superiority of the NE system over the DE system for predicting performance of growing pigs (Table 11.9).

**Table 11.9**  EFFECT OF ENERGY SYSTEM ON PREDICTION OF PERFORMANCE OF GROWING PIGS.

|  | *Dietary CP, %*[1] | | | *Dietary fat, %*[2] | |
|---|---|---|---|---|---|
|  | *17.8* | *15.5* | *13.6* | *1.8* | *8.2* |
| Feed:gain ratio: | | | | | |
| MJ DE/kg BW gain | 38.3[a] | 37.4[b] | 37.2[b] | 34.0[a] | 32.3[b] |
| MJ NE/kg BW gain | 27.6 | 27.5 | 27.6 | 24.4 | 24.0 |

[1]   After Quiniou *et al.* (1994); over the 30 to 100 kg BW period.
[2]   From unpublished data; over the 40 to 85 kg BW period.

## Practical consequences

The above sections have clearly demonstrated that the hierarchy between feeds is dependent on the energy system, the biggest differences being observed for ingredients whose chemical composition is quite different from that of standard diets. Results in least-cost formulation will therefore depend on the energy system (J. Noblet *et al.*, unpublished results). For instance, diets have lower protein contents when formulated on a NE concept than on a DE basis and a subsequent higher supplementation of synthetic amino acids. Fat inclusion levels are usually higher with the NE system. Within NE systems, diets formulated with the NEs or NEnl systems have lower starch contents than with the NEg systems. But when changing from DE (or ME) to NE systems, attention should be paid to the reduction of dietary protein and amino-acids levels with subsequent sublimiting available amino-acids supplies to the pigs. Consequently, adoption of an accurate protein evaluation system (available or digestible amino-acids) when a NE system is used is strongly recommended in order to adapt feed composition more precisely to requirements and growth potential of the pig.

According to methodological considerations, fractionation method of organic matter and accuracy of the equations (Noblet and Henry, 1993), the NEs, NEnl and NEj systems have limits for predicting performance of pigs and discriminating ingredients for their 'true' energy value. The NE prediction equations proposed by Noblet *et al.* (1994a) have been validated in growth trials, are applicable for both compound feeds and raw materials and can be used at all stages of pig production. In addition, they can be used directly from information available in feeding tables: DE or ME contents plus some chemical characteristics (ARC, NRC or INRA tables) for equations NEg4 and NEg7 (Table 11.6). For equation NEg2 based on digestible nutrient contents, values can be obtained from feeding tables (DLG, 1984; CVB, 1988), it can be assumed that starch is 100% digestible. Furthermore, in the case of ingredients providing mainly one specific nutrient (starch or fat), it is suggested that it is preferable to use equation NEg2. Finally, for free amino-acids, it is suggested that their NE value is 75% of their gross energy content.

Results presented in this review show that DCe is not constant for a given feed at all stages of pig production. Therefore, even if the efficiency of utilization of ME for NE can be considered as constant at all stages of pig production, different NE values should be used because of large differences in DE or digestible nutrient contents between stages. As previously mentioned, two 'model' stages are recommended: adult sows (whatever their feeding level) and 60 kg pigs as representative of the 30 to 100 kg BW period.

Another advantage of NE systems is that energy requirements of animals can be expressed on the same basis as energy values of feeds. Two alternatives can be used to transform energy requirements expressed on a DE basis to NE requirements. The simplest is to multiply daily DE recommendations by 0.71; this ratio corresponds to the mean NE:DE ratio measured by Noblet *et al.* (1994a). The most complex is to consider total energy requirements as the sum of NE requirements for maintenance and NE requirements for growth or for lactation (i.e., factorial approach). Under thermoneutral conditions, NE maintenance requirements (or FHP) average 750 kJ per kg $BW^{0.60}$ in growing pigs at a reduced level of activity (Noblet *et al.*, 1991 and 1994a) and 260 kJ per $kg^{0.75}$ in adult sows at zero level of physical activity (Noblet *et al.*, 1993c). Requirements for growth will be equal to the amount of retained energy, resulting from the addition of protein and fat energy retentions. Similarly, NE requirements for milk production can be predicted from the amount of milk energy (Noblet and Etienne, 1989) and by assuming similar values for $k_g$ and $k_l$. The calculation of NE requirements for thermoregulation is slightly more complicated since heat loss is partially used for thermoregulatory purposes; this aspect deserves further study. Energy requirements for physical activity should also be considered. Undoubtedly this approach, which is able to quantify energy requirements under all circumstances, will involve modelling techniques.

## Conclusions

The information reported in this review shows the limits and inadequacies of all energy systems. None of them is able to predict the «true» energy value of a diet and, subsequently, the performance of the pigs. However, NE systems should be preferred, especially for assessing the energy value of raw materials. But NE systems should be coupled with protein evaluation of feeds based on digestible (or available) amino acids levels.

It was commonly accepted that DE is a value which is specific of a feed and not affected by animal factors (ARC, 1981) while NE was dependent on both feed characteristics and animal factors. Our studies have demonstrated that the DE to NE step is mainly affected by feed characteristics and the GE to DE step is dependent on both feed chemical characteristics and animal factors. Therefore, from our point of view, the main limit for predicting correctly the energy value of a feed remains the estimation of its DE or digestible nutrient contents. Attention should be focussed on the effects of BW and interactions between BW and feed characteristics. In addition, since fibre is the main factor contributing to variations in digestive utilization of the diet, more emphasis should be given to routine techniques that identify the nutritional and physiological "quality" of dietary fibre.

The feed industry requires rapid and accurate prediction methods for evaluating the nutritional value of feeds. Measurements with animals have interest only in research or for calibration purposes. Use of values in feeding tables or predicted from equations give a reasonable estimate of the nutritional value of ingredients. However, available analytical procedures are often inadequate. Most of the progress in rapidity and reliability of nutritional value estimates of pig feeds will come from proposals concerning the application of *in vitro* methods or more sophisticated physico-chemical techniques. The main interest of studies on animals will then concern the applicability and the limits of such estimates.

## References

Agricultural Research Council (1981) *The Nutrient Requirements of Pigs*. Slough: Commonwealth Agricultural Bureaux

Batterham, E.S., (1990) Prediction of the dietary energy value of diets and raw materials for pigs. *Feedstuff Evaluation*. Eds J. Wiseman, D.J.A. Cole. p267-281. London; Butterworths.

Borggreve, G.J., Van Kempen, G.J.M., Cornelissen, J.P. and Grimbergen, A.H.M.(1975) The net energy content of pig feeds according to the Rostock formula. The value of starch in the diet. *Zeitschrift für Tierphysiologie, Tierernährung und Futtermittelkunde*, **34**, 199-204

Bourdon, D. (1986) Valeur nutritive des nouveaux tourteaux et graines entières de colza à basse teneur en glucosinolates pour le porc à l'engrais. *Journées de la Recherche Porcine en France*, **18**, 91-102

Chabeauti, E., Noblet, J. and Carré, B. (1991) Digestion of plant cell walls from four different sources in growing pigs. *Animal Feed Science and Technology*, **32**, 207-213

Central Veevoederbureau (1986) *Veevoedertabel. Gegevens over voederwarde, verteerbaarheid en samenstelling.* Lelystad: CVB

De Goey, L.W. and Ewan, R.C. (1975) Energy values of corn and oats for young swine. *Journal of Animal Science*, **40**, 1052-1057

Deutsche Landwirtschafts-Gesellschaft (1984) *Futterwertabellen für Schweine.* Frankfurt am Main: DLG-Verlag

Everts, H., Smits, B. and Jongbloed, A.W. (1986) Effect of crude fibre, feeding level and body weight on apparent digestibility of compound feeds by swine. *Netherlands Journal of Agricultural Science*, **34**, 501-503

Ewan, R.C. (1989) Predicting the energy utilization of diets and feed ingredients by pigs. In *Energy Metabolism of Farm Animals*, pp 215-218. Edited by Y. van der Honing and W.H. Close. Wageningen: Pudoc

Henry, Y, Vogt, H. and Zoiopoulos, P.E. (1988) Feed evaluation and nutritional requirements. III. 4. Pigs and Poultry. *Livestock Production Science*, **19**, 299-354

INRA (1989) *L'alimentation des Monogastriques (porc, lapin, volailles).* Paris: INRA [Feeding of non-ruminant livestock. Wiseman, J. 1990. Translated from the French and edited. Published by Butterworths, London].

Just, A. (1982a) The net energy value of crude fat for growth in pigs. *Livestock Production Science*, **9**, 501-509

Just, A. (1982b) The net energy value of crude (catabolized) protein for growth in pigs. *Livestock Production Science*, **9**, 349-360

Just, A. (1982c) The influence of crude fibre from cereals on the net energy value of diets for growth in pigs. *Livestock Production Science*, **9**, 569-580

Just, A. (1982d) The net energy value of balanced diets for growing pigs. *Livestock Production Science*, **8**, 541-555

Just, A., Jorgensen, H. and Fernandez, J.A. (1984) Prediction of metabolizable energy for pigs on the basis of crude nutrients in the feeds. *Livestock Production Science*, **11**, 105-128

Morgan, C.A., Whittemore, C.T., Phillips, Patricia and Crooks, P. (1987) The prediction of the energy value of compounded pig foods from chemical analysis. *Animal Feed Science and Technology*, **17**, 81-107

Noblet, J., Le Dividich, J. and Bikawa, T. (1985) Interaction between energy level in the diet and environmental temperature on the utilization of energy in growing pigs. *Journal of Animal Science*, **61**, 452-459

Noblet, J. and Etienne, M. (1989) Estimation of sow milk nutrient output. *Journal of Animal Science*, **67**, 3352-3359

Noblet, J., Fortune, H., Dubois, S. and Henry, Y. (1989a) *Nouvelles bases d'estimation des teneurs en énergie digestible, métabolisable et nette des aliments pour le porc*. Paris: INRA

Noblet, J., Dourmad, J.Y., Le Dividich, J. and Dubois, S. (1989b) Effect of ambient temperature and addition of straw or alfafa in the diet on energy metabolism in pregnant sows. *Livestock Production Science*, **21**, 309-324

Noblet, J., Dourmad, J.Y. and Etienne, M. (1990) Energy utilization in pregnant and lactating sows: modelling of energy requirements. *Journal of Animal Science*, **68**, 562-572

Noblet, J. and Henry, Y. (1991) Energy evaluation systems for pig diets. In *Manipulating Pig Production III*, pp 87-110. edited by E.S. Batterham. Attwood: Australasian Pig Science Association

Noblet, J., Karege, C. and Dubois, S. (1991) Influence of growth potential on energy requirements for maintenance in growing pigs. In *Energy Metabolism of Farm Animals*, pp 107-110. Edited by C. Wenk and M. Boessinger. Zurich: ETH

Noblet, J. and Henry, Y. (1993) Energy evaluation systems for pig diets: a review. *Livestock Production Science*, **36**, 121-141

Noblet, J., Shi, X.S. (1993) Comparative digestibility of energy and nutrients in growing pigs fed ad libitum and adult sows fed at maintenance. *Livestock Production Science*, **34**, 137-152.

Noblet, J., Shi, X.S., Karege, C. and Dubois, S. (1993a) Effets du type sexuel, du niveau d'alimentation, du poids vif et du stade physiologique sur l'utilisation digestive de l'énergie et des nutriments chez le porc; interactions avec la composition du régime. *Journées de la Recherche Porcine en France*, **25**, 165-180

Noblet, J., Fortune, H., Dupire, C. and Dubois, S. (1993b) Digestible, metabolizable and net energy values of 13 feedstuffs for growing poigs: effect of energy system. *Animal Feed Science and Technology*, **42**, 131-149

Noblet, J., Shi, X.S. and Dubois, S. (1993c) Metabolic utilization of dietary energy and nutrients for maintenance energy requirements in sows: basis for a net energy system. *British Journal of Nutrition*, **70**, 407-419

Noblet, J. and Perez, J.M. (1993) Prediction of digestibility of nutrients and energy values of pig diets from chemical analysis. *Journal of Animal Science*, **71**, 3389-3398

Noblet, J. and Shi, X.S. (1994) Effect of body weight on digestive utilization of energy and nutrients of ingredients and diets in pigs. *Livestock Production Science*, **37**, 323-338

Noblet, J., Fortune, H., Shi, X.S. and Dubois, S. (1994a) Prediction of net energy value of feeds for growing pigs. *Journal of Animal Science*, **72**, 344-354

Noblet, J., Shi, X.S. and Dubois, S. (1994b) Effect of body weight on net energy value of feeds for growing pigs. *Journal of Animal Science*, **72**, 648-657

Noblet, J., Shi, X.S., Fortune, H., Dubois, S., Lechevestrier, Y., Corniaux, C., Sauvant, D. and Henry, Y. (1994c) Teneur en énergie nette des aliments chez le porc: mesure, prédiction et validation aux différents stades de sa vie. *Journées de la Recherche Porcine en France*, **26**, 235-250

National Research Council (1988) *Nutrient Requirements of Swine, 9th edn.* Washington DC: National Academic Press

Perez, J.M., Ramoelintsalama, B. and Bourdon, D. (1980) Prédiction de la valeur énergétique de l'orge pour le porc à partir des teneurs en constituants membranaires. *Journées de la Recherche Porcine en France*, **12**, 273-285

Perez, J.M., Ramihone, R. and Henry, Y. (1984) *Prédiction de la valeur énergétique des aliments composés destinés au porc: étude expérimentale.* Paris: INRA

Perez, J.M., Bourdon, D., Baudet, J.J. and Evrard, J. (1986) Prévision de la valeur énergétique des tourteaux de tournesol à partir de leurs teneurs en constituants pariétaux. *Journées de la Recherche Porcine en France*, **18**, 35-42

Quiniou, N., Dourmad, J.Y., Henry, Y., Bourdon, D. and Guillou, D. (1994) Influence du potentiel de croissance et du taux protéique du régime sur les performances et les rejets azotés des porcs en croissance-finition alimentés à volonté. *Journées de la Recherche Porcine en France*, **26**, 91-96

Roth, F.X. and Kirchgessner, M. (1984) Verdaulichkeit der Energie und Rohnährstoffe beim Schwein in Abhängigkeit von Fütterungsniveau und Lebendgewicht. *Zeitschrift für Tierphysiologie, Tierernährung und Futtermittelkunde*, **51**, 79-87

Schiemann, R., Nehring, K., Hoffmann, L., Jentsch, W. and Chudy, A. (1972). *Energetische Futterbevertung und Energienormen.* Berlin: VEB Deutscher Landwirtschatsverlag

Shi, X.S. and Noblet, J. (1993) Digestible and metabolizable energy values of ten ingredients in growing pigs fed ad libitum and sows fed at maintenance level; comparative contribution of the hindgut. *Animal Feed Science and Technology*, **42**, 223-236

**12**

## ENERGY-AMINO ACID INTERACTIONS IN MODERN PIG GENOTYPES

T.A. Van Lunen
*Atlantic Veterinary College, University of Prince Edward Island, Charlottetown, Prince Edward Island, C1A 4P3, Canada*

and

D.J.A. Cole[1]
*University of Nottingham, Sutton Bonington Campus, Loughborough, Leicestershire, LE12, 5RD, UK*

### Introduction

In considering the nutrient requirements of growing pigs, the measurement of growth which is most appropriate must be selected and the factors which affect that growth identified.

The deposition of protein is of prime importance to growth performance as it represents a useful method for measuring the rate of lean tissue growth. Increases in protein deposition are usually inversely correlated with lipid deposition (Fowler, 1978) and slightly positively correlated with water content (Stranks *et al.*, 1988). Thus improvements in protein deposition will usually be at the expense of lipid with a slight positive effect on water content. The rate of protein deposition (PDR) can be defined as the increase in the total amount of protein in the body of the pig.

### Factors affecting growth

AGE/LIVE WEIGHT

It may appear at first glance that age and live weight are synonymous; older animals being heavier and therefore larger than younger animals. Although this is generally true, the genetic growth potential of the animal, in this case the pig, can confound

---

[1]Current address:

Nottingham Nutrition International, Potters Lane, East Leake, Loughborough, Leicestershire, LE12 6NQ, UK

this assumption so that younger pigs in one genetic group can be heavier and larger than older pigs in another group.  It is therefore important to differentiate between age and weight when considering their effects on growth.

Within genotype and sex, provided adequate levels of nutrients are supplied, it is inevitable that PDR will increase during early growth, reach a peak, and then decrease as the animal increases in age and live weight (Carr, Boorman and Cole, 1977; Yen, Cole and Lewis, 1986b).  The decline in PDR as the animal approaches maturity is accompanied by an increase in lipid deposition rate (LDR) (Whittemore, 1993).  According to Campbell (1990) there are currently two schools of thought concerning the form of the response curve of PDR to live weight or age.  The first is a quadratic response with maximum PDR occurring at 70 to 90 kg live weight in a relatively sharp peak.  The second assumes that PDR increases rapidly during early life resulting in maximum PDR occurring at 20 to 40 kg live weight and maintaining a plateau until approximately 120 kg live weight.  Dunkin, Black and James (1986) and Dunkin and Black (1985) suggested that the potential rate of nitrogen deposition increases with live weight up to 70 to 80 kg, after which it begins to decline.  On the other hand, Whittemore (1986) suggested that PDR is largely unaffected by live weight.  Whittemore, Tullis and Emmans (1988) reported that PDR reached its maximum at 70 kg live weight, but that a general plateau of PDR existed from 45 to 125 kg live weight.  This plateau of PDR over a wide weight range has also been reported by Kielanowski (1969), Stranks *et al*. (1988) and Whittemore (1993).

Whittemore and Fawcett (1976) proposed that potential protein deposition is limited by appetite in young pigs, but is restricted to a point below maximum voluntary feed intake in older pigs.  Work by Campbell *et al*. (1990) has indicated that in genetically superior pigs maximum protein deposition is limited by appetite, even in growing-finishing pigs (Figure 12.1).  In general, genotypes with higher rates of protein deposition are those with greater mature weight (Ferguson and Gous, 1993) and, since these superior lines of pigs reach a specific weight at a younger age, it is apparent that physiological age may be more important in the control of protein deposition than live weight.  The differing results reported in the literature regarding weight at maximum PDR as well as the shape of the response curve of PDR to live weight may be explained by differences in genetic make up.  From these results, however, it is still apparent that as the pig approaches maturity its potential for protein deposition decreases.

SEX

Sex of the animal has a strong effect on maximum protein deposition potential and on the protein deposition response to nutrient levels in pigs above 50 kg liveweight with boars having the highest response, castrates the lowest response and gilts being intermediate (Prescott and Lamming, 1967; Newell and Bowland, 1972; Taverner, Campbell and King, 1977; Cresswell *et al.*, 1975; Williams *et al.*, 1984; Yen, Cole and Lewis, 1986a). Figure 12.2 illustrates the PDR response of boars, gilts, and castrates, as well as different genotypes to energy intake. The figure demonstrates that higher PDR potential allows the animal to respond to higher energy intake before a lean growth plateau is reached.

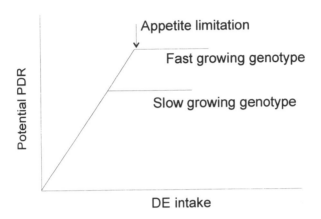

**Figure 12.1** Potential PDR of two genotypes of pigs in relation to DE intake

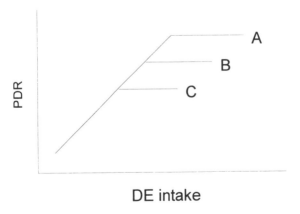

**Figure 12.2** Potential PDR of pigs with different growth capacities in relation to DE intake
*A= boars or highly selected genotype, B=gilts or average genotype, C=castrated male or slow genotype*

It appears that intrinsic factors present in the different sexes determine protein deposition. The major differences between them are levels of sex hormones, thus implying that the endocrine system has a role to play in protein deposition. This is demonstrated by the results of work with exogenous growth promotants such as recombinant porcine somatotropin (pST) where effects of pST on protein deposition were of similar magnitude for castrates and gilts while the relative protein deposition advantage of the boars was maintained (Kanis *et al.* 1990).


GENOTYPE

Breed differences are well known to affect body composition and growth. In pigs, for example, the Pietrain breed can have up to 100 g/kg live weight more muscle than the Large White breed at the same live weight (Lindsay, 1983). This is a reflection of the amount of lipid present in the body as much as it is the total amount of lean. Breeds of animals with a higher proportion of lean tissue at a given live weight are those which mature at a larger body size (Whittemore, 1993). Thus comparison of two breeds (ie Pietrain *vs* Large White) of equal weights involves comparison of animals with differing physiological age. In other words, the physiologically younger animal is the one with less lipid per unit body weight since lipid is deposited in greater quantities than lean as the animal approaches maturity.

Siebrits and Kemm (1982) reported differences in maximum potential for lean tissue deposition between obese and lean strains of pigs. This suggests that genotype not only affects lean deposition by altering physiological age at a given weight, but also the potential maximum level of lean deposition is affected. Work reported by Ellis *et al.* (1983) and Henderson *et al.* (1983) indicated that selected pigs had a faster lean deposition rate than unselected pigs under both *ad libitum* and restricted feeding conditions. These studies also indicated that when nutrients were not limited, the selected pigs had a steeper linear response of lean deposition to nutrient intake than unselected pigs.

A classic study reported by Campbell and Taverner (1985) compared two strains of pigs (A and B) fed graded energy levels in protein adequate diets. Figure 12.2 is also used to illustrate the protein deposition responses of different genotypes of pigs. This characteristic response is manifested as a linear PDR increase with increasing dietary energy followed by a plateau. In the Campbell and Taverner (1985) trial, Strain A pigs reached a PDR plateau of 132 g/d at an energy intake of 32.5 MJ DE per day while Strain B pigs increased PDR in response to DE intake until maximum appetite was reached (39 MJ DE/day). At this intake level PDR reached 186 g/d. These results indicated that the maximum potential PDR for Strain B pigs was at, or somewhere above, the limit of voluntary feed consumption, while Strain A pigs had a genetic PDR response potential considerably below the level of appetite. This appetite difference can also exist for boars as compared to gilts and castrates as discussed earlier.

Despite the higher potential performance of highly selected pigs, it also appears that such animals are more sensitive to nutritional deficiencies than lower PDR potential animals. Campbell (1988) reported that the level of dietary lysine required to support maximum protein deposition in Strain B pigs was 18 to 20% higher than the requirement for Strain A pigs. Rao and McCracken (1991) reported that the response of improved genotypes in protein deposition increases with increasing ME intake above normal appetite levels and any restriction in energy intake would have a greater negative impact on protein deposition than in unimproved pigs. It was also reported that the maintenance requirement for improved genotypes was greater than for unimproved ones.

Work reported by Cole and Chadd (1989) and Chadd, Cole and Walters (1993) suggested that faster growing genotypes have a lower appetite than conventional pigs. As a result, the maximum PDR may be hindered in these animals if appetite is the first limiting factor to maximum protein deposition.

From the above it is clear that genotype has a strong influence on PDR in pigs. Genotypes with the potential for high PDR have a higher dietary requirement for protein and, perhaps more importantly, energy. Any restriction in intake of either of these dietary components can have a strong negative impact on protein deposition.

## NUTRITION

### Feed supply

According to Whittemore (1993) control of the feed supply to the pig is the most biologically, financially and managementally effective way of optimizing pig performance. Assuming that the diet provides an adequate balance of nutrients to meet the maintenance and production requirements of the animal, the amount consumed on a daily basis will have a strong impact on growth performance and carcass quality.

Before the development of modern genotypes capable of fast, lean growth, it was generally recognized that maximum lean growth was achieved at a point below the level of appetite, especially in the finishing phase. In other words, increasing feed intake from a slightly restricted level to *ad libitum* would result in increased lipid deposition with no increase in protein deposition. As a result, conventional feeding practices utilized restricted feeding to reduce growth rate and maximize lean content and carcass quality. Most research conducted to evaluate the effects of restricting feed intake has been done using conventional genotypes and it is well established that, for such pigs, restricting feed intake to a level from 85 to 95% of *ad libitum* will slightly reduce growth rate and improve feed conversion. According to English *et al.* (1988) for each 0.1 kg decrease in daily feed consumption, pigs from 25 to 90 kg will experience a decrease in growth rate of 0.035 kg/day. Campbell, Taverner and Curic (1985) reported an optimal feed conversion in pigs from 48 to 90 kg when feed was restricted by 15 to 20% of *ad libitum*. In this trial a 10% restriction in feed allocation resulted in a 10% improvement in feed conversion in boars and a 6% improvement in gilts. In commercial practice feed restriction may further improve feed efficiency by reducing feed wastage.

The primary purpose of restricting feed intake in growing pigs is to reduce fat content in the carcass. The effects of restricted feeding on deposition of lipid and lean are illustrated in Figure 12.3. In normal growth, as feeding level increases, both fat and lean deposition increase at similar rates until the maximum genetic potential for lean deposition is reached. Beyond this point, further increases in feeding level will result in large increases in fat deposition with very little, if any, accompanying increase in lean deposition.

For every 0.1 kg reduction in daily feed intake below *ad libitum*, backfat thickness in pigs slaughtered at 90 kg will be reduced by 0.52 mm (English *et al.,* 1988) . Work reported by Kanis (1988) suggested similar trends with a 0.4 mm reduction in backfat thickness per 0.1 kg reduction in feed consumption for pigs slaughtered at 108 kg. The latter work also suggested that for each 0.1 kg reduction in feed intake, was a resultant 5% improvement in carcass lean tissue and a 5.5% reduction in carcass fat.

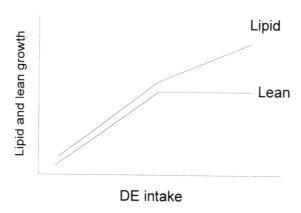

**Figure 12.3** Lipid and lean growth in response to feed supply

The development of genotypes which exhibit rapid, lean growth has altered previously accepted theories regarding nutrient requirements of pigs. Such pigs grow at rates exceeding 1.0 kg/day with feed conversion ratios approaching 2:1. Figure 12.4 illustrates the difference in response to level of feeding of conventional pigs and fast growing genotypes. In the fast growing pig the potential for lean deposition is much greater than in the conventional pig. Until feed intake is sufficient to maximize lean tissue growth, the pig will lay down a minimum of fat during growth (Whittemore, 1993). Once the level of feed required to maximize lean deposition is reached, fat deposition rate will increase rapidly with increases in feed consumption. Since highly selected genotypes and boars can increase PDR with increasing energy intake up to their appetite limit, feed restriction is not recommended for these animals.

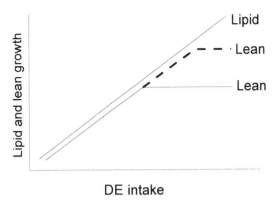

**Figure 12.4** Lipid and lean growth in response to feed supply of conventional and modern genotypes
— Conventional genotype    - - - Modern genotype

*Voluntary feed intake*

With the shift to *ad libitum* feeding (away from restricted and allowance feeding), the level of feed intake achieved and the factors which affect this assume considerable importance.

In order to establish values for contemporary pigs, Chadd, Cole and Walters (1993) developed the voluntary feed intake equation:

$$DE(MJ) = 5.00W^{0.44} \tag{1}$$

It is clear that voluntary feed intake has declined with the development of more modern genotypes, and that the highly selected pigs have lower appetite levels than more conventional genotypes (Figure 12.5). The depression in appetite of modern pigs compared with the slower growing genotypes (for example 30 years ago) emphasizes the problems associated with DE intake for the achievement of the genetic potential for high rates of lean gain. Increases in PDR, and thus lean gain, have not been associated with increases in lipid deposition. Thus feed or energy restriction to reduce lipid deposition, even in gilts and castrates, should not be practised with modern genotypes during the growing period.

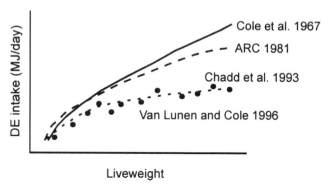

Figure 12.5 Changes in voluntary digestible energy intake with selection of modern genotypes

## PROTEIN

### *Ideal protein*

The most important single factor affecting the efficiency of protein utilization for production of meat or eggs is the dietary balance of amino acids. According to Liebig's "Law of Minimums" the undersupply of a single essential amino acid will inhibit the responses to those in adequate supply. As a result, the balance of essential amino acids in the diet of the pig is essential for the efficient use of protein in order to maximize growth.

In rationalizing the amino acid requirements of pigs, Cole (1978) suggested the adoption of the concept of an ideal protein. In order to examine the pattern of dietary amino acids in non-ruminant animals, the ideal protein gives a simple and effective approach. The development of an ideal protein in pigs has received much attention in recent years. The work of Cole (1978) and Fuller *et al.* (1979) was combined and used by ARC (1981) as the basis for protein requirements in pigs. It had been suggested that the major differences between pigs of different classes (i.e. breed, sex and live weight) would be the amount of protein that they require according to their potential for lean meat deposition. The relative amounts of the different essential amino acids needed for 1 g lean or protein deposition would be the same in each case. Thus it was suggested that it should be possible to establish an optimum balance of essential amino acids which, when supplied with sufficient protein for the synthesis of non-essential amino acids, would constitute the ideal protein. Pigs of different classes would require different amounts of the ideal protein but the quality would be the same in each case (Cole, 1978).

While such a concept has probably been best developed in pigs, it is equally applicable to other species. It also interesting to note that the amino acid pattern of the diet appears to have no effect on the amino acid makeup of the pig (Kemm, Siebrits and Barnes, 1990) further emphasizing that the dietary amino acid requirement is a function of the amino acid composition of the animal.

Lysine was chosen as the basis for the establishment of the essential amino acid pattern as it is required in large amounts for protein deposition and is almost always the first limiting amino acid in cereal based diets. According to Batterham (1994) the development of the 'ideal protein' has simplified dietary formulations and facilitated the use of lysine/DE ratios in the establishment of appropriate dietary protein/energy balances.

Work on the establishment of precise numbers for the ideal protein was particularly active in the 1970's. On the basis of this, suggestions were made for the balance of essential amino acids in the ideal protein (Table 12.1).

**Table 12.1** THE OPTIMUM BALANCE OF ESSENTIAL AMINO ACIDS IN THE IDEAL PROTEIN FOR PIGS (RELATIVE TO LYSINE = 100)

| | Cole | Fuller et al. | ARC | Fuller and Wang (1987) | | | Proposed balance |
| --- | --- | --- | --- | --- | --- | --- | --- |
| | | | | | Tissue | | |
| | 1978 | 1979 | 1981 | Maintenance | deposition | Both | |
| Lysine | 100 | 100 | 100 | 100 | 100 | 100 | 100 |
| Methionine + cystine | 50 | 53 | 50 | 150 | 53 | 56 | 50-55 |
| Tryptophan | 18 | 12 | 15 | 29 | 18 | 19 | 18 |
| Threonine | 60 | 56 | 50 | 142 | 69 | 72 | 65-67 |
| Leucine | 100 | 83 | 100 | 71 | 115 | 113 | 100 |
| Valine | 70 | 63 | 70 | 54 | 76 | 75 | 70 |
| Isoleucine | 50 | 50 | 55 | 46 | 63 | 63 | 50 |
| Phenylalanine + tyrosine | 100 | 96 | 96 | 125 | 124 | 123 | 100 |
| Histidine | 40 | 31.5 | - | - | - | - | - |

As the knowledge of amino acid requirements increases, so the values used in the original work need to be questioned. However, the original balances have proved to be particularly durable. Nevertheless, some amino acids are worthy of attention, namely threonine and methionine + cystine.

In the original balance, a value of dietary threonine of 60% of the lysine level received substantial support. More recently evidence has been accumulating that pigs can respond to threonine levels of up to 65 to 67% of the dietary lysine (Cole, 1993 *unpublished data*).

While there was considerable variation in the estimates of requirements for methionine + cystine, a value of 50-55% of the dietary lysine supply has generally been adopted. Work at the University of Nottingham (Cole, 1993 *unpublished data*) has consistently supported a value of 50%. The work suggested that higher requirement values for methionine + cystine may well occur when cystine was providing a substantial part of their joint requirements (Figure 12.6). As a result it was recommended that dietary methionine should not be allowed to fall below 0.25%.

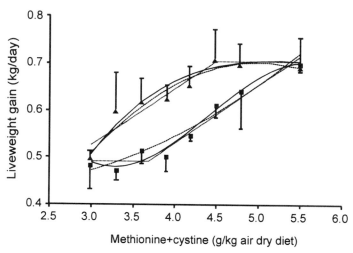

**Figure 12.6** The response of growing pigs to methionine and cystine supply (▲ 75% methionine, 25% cystine; ■ 25% methionine, 75% cystine)

In the original description of the ideal protein (Cole, 1978; ARC, 1981) it had been calculated that as maintenence requirements were generally only 1 to 3% of production (Baker *et al.*, 1966; Fuller and Wang, 1987) then for practical purposes a single value could be used for pigs of different weights. Recently, values have been presented (Table 12.2) suggesting that the differing needs of maintenance and production result in substantially higher values for some amino acids at higher live weights. The values for methionine + cystine rise from 60% of lysine for 5 to 20 kg liveweight pigs to 70% of lysine for 50 to 100 kg live weight pigs. Such recommendations need to be questioned from two standpoints. Firstly, the absolute levels (e.g. 60% for the young pig) are much higher than the literature suggests. Second, the very low requirement for maintenance relative to production makes a shift of 10% difficult to attain in theory or practice. Calculation of the influence of live weight can be made from the data obtained in the University of Nottingham work and applying the values of Fuller and Wang (1987) (Table 12.3). Such calculations support the empirical data that large increases in the ratio of some individual amino acids as live weight increases are not appropriate.

**Table 12.2**   'IDEAL' PATTERNS OF ESSENTIAL AMINO ACIDS FOR GROWING PIGS AT THREE LIVE WEIGHTS (RELATIVE TO LYSINE = 100)

|                             | *5 to 20 kg* | *20 to 50 kg* | *50 to 100 kg* |
|-----------------------------|--------------|---------------|----------------|
| Lysine                      | 100          | 100           | 100            |
| Methionine + cystine        | 60           | 62.5          | 70             |
| Tryptophan                  | 18           | 19            | 20             |
| Threonine                   | 65           | 67            | 70             |
| Leucine                     | 100          | 100           | 100            |
| Valine                      | 68           | 68            | 68             |
| Isoleucine                  | 60           | 60            | 60             |
| Phenylalanine + tyrosine    | 95           | 95            | 95             |
| Histidine                   | 32           | 32            | 32             |

From Baker *et al.* (1993)

**Table 12.3** INFLUENCE OF LIVE WEIGHT ON ESSENTIAL AMINO ACID BALANCE IN THE IDEAL PROTEIN (LYSINE=100%) BASED ON THE VALUES OF FULLER AND WANG (1987).

| *Livewight (kg)*        | *20*  | *70*  |
|-------------------------|-------|-------|
| Lysine                  | 100   | 100   |
| Methionine + cystine    | 55.4  | 56.6  |
| Threonine               | 71.0  | 72.3  |
| Tryptophan              | 18.4  | 18.6  |

When the ideal protein was introduced it was done so on the basis of total amino acids. This was a deliberate attempt to remove as much error as possible; in this case the variable results from different 'availability' techniques.

Ileal digestibility is now used widely as a measure of availability. Some recent work (e.g. Wang and Fuller, 1990) uses ileal digestible amino acids as the basis for calculating the ideal protein. As the individual amino acids are absorbed to different extents, this will also cause a shift in the ratios used.

In many cases ileal digestibility is a good measure of availability (e.g. Tanksley and Knabe, 1984). However, care needs to be taken in using such an application. For example, it has been shown that lysine retention as a proportion of ileal digestible lysine intake was influenced by dietary lysine concentration (Batterham *et al.*, 1990).

A further example is the case of heat-damaged fish meals. Work at the University of Nottingham (Table 12.4) has shown a reduction in both faecal and ileal digestibility of lysine (Wiseman *et al.*, 1991). However, formulation of diets on the basis of either type of digestibility, while giving better performance, did not completely account for problems of heat damage. Pigs given diets containing heat-damaged fish meal, based on total, faecal or ileal digestible amino acids, grew at 64%, 83% and 87% respectively of the rate of pigs fed diets containing undamaged fish meal.

**Table 12.4** ILEAL AND FAECAL DIGESTIBILITIES IN PIGS OF LYSINE IN FISHMEALS HEATED TO DIFFERENT TEMPERATURES (WISEMAN ET AL., 1991)

|  | *Ileal digestibility of lysine* | *Faecal digestibility of lysine* |
|---|---|---|
| Fishmeal |  |  |
| untreated | 0.92 | 0.99 |
| 3h @ 130° | 0.90 | 0.95 |
| 1.25h @ 160° | 0.84 | 0.91 |

Values for the ratio of ileal digestible essential amino acids are given in Table 12.5. These are based on published work and may not represent the perfect balance.

**Table 12.5** RATIOS OF THE ILEAL DIGESTIBLE ESSENTIAL AMINO ACIDS IN THE IDEAL PROTEIN OF PIGS (25 - 55 KG LIVEWEIGHT) USED BY YEN, COLE AND LEWIS (1986A, 1986B) AND WANG & FULLER (1990)

| *Amino acid* | *Yen et al. (1986a)* | *Yen et al. (1986b)* | *Wang & Fuller (1990)* |
|---|---|---|---|
| Lysine | 100 | 100 | 100 |
| Methionine | 37 | 39 | - |
| Methionine + cystine | 52 | 58 | 60 |
| Threonine | 64 | 67 | 66 |
| Tryptophan | 19 | 21 | 18.5 |
| Isoleucine | 73 | 76 | 60 |
| Leucine | 130 | 140 | 111 |
| Phenylalanine + tyrosine | 86 | 95 | 120 |
| Histidine | 43 | 46 | - |
| Valine | 90 | 97 | 75 |

Availability can be considered in another sense. It has been suggested that synthetic amino acids need to be in phase, at the metabolic sites, with the protein bound amino acids. To ensure this, it has been suggested that it is necessary to feed more frequently than once per day when using high levels of crystalline lysine (Batterham, 1974; Batterham and O'Neill, 1978) and threonine (Cole, 1992). However in modern production systems, using *ad libitum* feeding, such problems are unlikely to be encountered.

### Protein requirement for maintenance and growth

Once the ideal pattern of amino acids within dietary protein has been established, the protein requirement, as a whole, for maintenance and growth can be discussed.

Protein is required in the body to replace losses due to protein turnover and for the production of enzymes and gut secretions and for the replacement of intestinal epithelial cells and skin (maintenance) as well as for lean tissue growth.

Protein requirements not only depend upon the amino acid pattern of the diet, but also the digestibility of the protein within the feedstuffs provided to the animal. As protein digestibility varies widely from one feedstuff to the next, protein requirements are best expressed in terms of digestible protein.

Whittemore (1993) reported the ideal protein requirement for maintenance ($IP_m$) in terms of body mass (W) and protein mass (Pt) as:

$$IP_m = 0.0013W^{0.75} \tag{2}$$

$$IP_m = 0.0040Pt \tag{3}$$

These equations assume that protein is required for maintenance in relation to the body mass or, more accurately, the protein mass of the body in order to supply protein for the functions outlined above.

Assuming that digestibility of protein within the diet has been determined, the estimation of dietary protein required for growth or protein deposition is a simple one where the requirement equals the rate of deposition (Whittemore, 1993). Therefore, the total protein requirement for maintenance and growth is the protein deposition rate plus the maintenance requirement as calculated above.

### ENERGY

Gut capacity of the pig from 20 to 50 kg live weight is 1.8 to 2.0 kg/d, consequently there is a dietary requirement of 14.5 to 15 MJ/kg DE in the diet in order to achieve the optimum DE intake of 30 to 32 MJ/d (Campbell 1990). Earlier work reported by Campbell *et al.* (1975) gave a slightly higher value of 15.3 MJ DE/d for young pigs. It has also been reported that slower growing pigs from 50 to 90 kg live weight require 32 MJ DE/d while faster growing pigs require 40 MJ DE/d to fuel the process of protein deposition (Campbell, 1990). Furthermore it has been suggested that fast growing pigs experienced a 6 g decrease in PDR for every 1 MJ reduction in DE intake below the optimum (Rao and McCracken, 1991). These values are helpful in getting an overview of the energy requirements of the growing pig, however more detailed estimates are required to improve accuracy and to provide an approach which is universal to all genotypes and growth stages of the pig.

Dietary energy is required to fuel the various growth and metabolic processes in the body. In general, the energy requirement of the pig, as in other animals, can be considered in two parts: the energy requirement for maintenance, and the energy requirement for growth. Within the growth requirement, separate requirements exist for protein and lipid deposition.

The maintenance requirement ($ME_m$) can be defined as the ME required for energy equilibrium (Stranks *et al.*, 1988). The ARC (1981) equation for growing pigs based on live weight (W):

$$ME_m \ (MJ) = 0.719W^{0.63} \tag{4}$$

appears accurately to predict $ME_m$ (Stranks *et al.*, 1988).

It has been suggested that the maintenance energy requirement for fast growing lean pigs was different from that of conventional pigs (Rao and McCracken, 1991) giving the equation:

$$ME_m \ (MJ) = 0.982/W^{0.63} \tag{5}$$

Further work is required to estimate the energy requirement for protein and lipid deposition. However it appears that protein requires of the order of 44 to 54 MJ ME/kg deposited while lipid deposition requires 53 MJ ME/kg deposited (Whittemore, 1993; Stranks *et al.*, 1988).

### Protein/energy interactions

Energy in excess of the needs of maintenance and the growth process is stored as fat within the body depots. Dietary protein is required as the substrate (both amino acids and non-essential nitrogen) for lean tissue growth and its associated protein turnover. Excess protein is deaminated and the protein portion excreted in the urine. This process of deamination and elimination has an energy cost associated with it. In order to maximize lean gain and minimize lipid gain, diets must be formulated with appropriate balances of protein and energy.

It is clear that lean tissue growth is affected by both protein and energy intake, each via its own mechanism (Campbell, 1988) resulting in protein and energy dependent phases of lean tissue growth. Under conditions of dietary protein deficiency, lean tissue growth responds to increasing protein intake. Similarly, under conditions of energy deficiency, lean tissue growth will increase with increasing energy intake. The objective, therefore, is to formulate diets which supply balanced proportions of protein and energy in order to maximize lean gain. According to Batterham (1994) this balance is affected by a number of factors such as growth phase, sex, environment and genotype. These factors all relate to the capacity for lean growth and therefore the balance of protein and energy can be discussed in these terms.

The relationship between protein and energy in the diet is most commonly referred to as the lysine/DE ratio where lysine is a reference for the 'ideal protein' and the values are expressed in terms of grams of lysine per MJ DE.

Many reports exist in the literature attempting to identify the optimum lysine/DE ratio for pigs at various stages of growth (Batterham, Giles and Dettmann, 1985; Campbell, Taverner and Curic, 1985; Giles, Batterham and Dettmann, 1986; Fuller *et al.*, 1986; Yen, Cole and Lewis, 1986a, 1986b; Campbell, Taverner and Curic, 1988; Campbell ,1990; Rao and McCracken, 1990; Cole, 1992; Nam and Aherne, 1994). Some of this work is summarized in Table 12.6.

### *Linear plateau vs curvilinear responses to lysine/DE ratios*

Most work associated with responses of protein intake to growth rate or PDR, such as Campbell (1988), has assumed that a plateau is reached when $DLWG_{max}$ or $PDR_{max}$ is achieved. This approach would result in a lower optimum lysine/DE ratio than a curvilinear (i.e. quadratic) response curve. The results of an experiment recently conducted at the University of Nottingham (Van Lunen and Cole, 1994) using highly selected pigs from 25 to 90 kg illustrate that this plateau does not exist, but rather DLWG, PDR and LDR decrease as the optimum lysine/DE ratio is exceeded. This was especially true for gilts and castrates, but may also have existed for boars. As a result, a curvilinear rather than linear/plateau approach to describing the relationship of protein to energy intake to growth or protein deposition appears more valid. Figure 12.7 illustrates both the linear-plateau and quadratic fit to the data of that trial.

**Table 12.6** COMPARISON OF OPTIMUM LYSINE/DE RATIOS (G/MJ) AS REPORTED
IN THE LITERATURE

| Sex | Body weight (kg) | Growth rate (kg/day) | Lysine/DE (g/MJ) | Source |
|---|---|---|---|---|
| Males | 20-50 | 0.79 | 0.69 | 1 |
| | 50-90 | 0.84 | 0.58 | 1 |
| Females | 20-50 | 0.74 | 0.62 | 1 |
| | 50-90 | 0.80 | 0.45 | 1 |
| Females | 18-65 | 0.67 | 0.81 | 2 |
| Castrates | | | (low intake) | |
| Males | 20-50 | 0.76 | 0.73 | 3 |
| | 50-90 | 0.78 | 0.59 | 3 |
| Males | 20-50 | 0.80 | 0.82 | 4 |
| | 50-90 | 0.95 | 0.62 | 4 |
| Castrates | 20-50 | 0.76 | 0.73 | 4 |
| | 50-90 | 0.82 | 0.54 | 4 |
| Gilts | 20-50 | 0.74 | 0.75 | 4 |
| | 50-90 | 0.86 | 0.51 | 4 |
| Males | 20-50 | 0.76 | 0.75 | 5 |
| | 50-90 | 0.83 | 0.51 | 5 |
| Females | 20-50 | 0.70 | 0.71 | 5 |
| | 50-90 | 0.78 | 0.44 | 5 |
| Males | 33-55 | 1.05 | 0.80 | 6 |
| | 55-88 | | | |
| Males | 20-50 | 0.40-0.77 | 0.82 | 7 |
| Males | 20-50 | 0.85 | 0.72 | 8 |

From Close (1994a)
1. Batterham, Giles and Dettmann (1985)
2. Fuller *et al.* (1986)
3. Giles, Batterham and Dettmann (1986)
5. Campbell, Taverner and Curic (1988)
4. Yen, Cole and Lewis (1986a, 1986b)
6. Rao and McCracken (1990)
7. Campbell, Taverner and Curic (1986)
8. Chiba, Lewis and Peo (1991)

   Support for a reduction in growth rate above an optimum lysine/DE ratio comes
from Cooke, Lodge and Lewis, 1972; Holmes, Carr and Pearson, 1980 and
Campbell, Taverner and Curic, 1984. In all those reports the pigs with the lowest

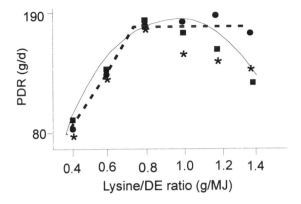

**Figure 12.7** PDR response to lysine/DE ratio of boars, gilts and castrates from 25 to 90 kg liveweight.

PDR potential were the first to exhibit a negative response to increases of lysine/DE ratio above the optimum.

The most likely cause of this negative effect is reduction in net energy of the diet due to the deamination of excess protein. As deamination is an inefficient process, less energy is available for PDR and growth as well as for lipid deposition. As energy intake is known to be a major limiting factor in the achievement of $PDR_{max}$ in fast growing genotypes, any reduction in net energy of the diet will negatively affect growth, PDR and LDR. Figure 12.8 illustrates the effect of increasing lysine/DE ratio on body lipid content.

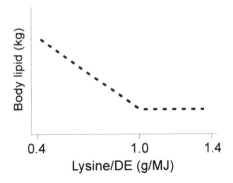

**Figure 12.8** Lipid content of growing pigs fed lysine/DE ratios from 0.4 to 1.4 g/MJ

The development of modern genotypes may help us more fully understand the energy and protein requirements of all growing pigs. From the results of the trials conducted at the University of Nottingham, it appears that modern genotypes selected for fast, lean growth differ in many aspects to slower growing genotypes. Of obvious importance is the PDR potential of these genotypes and the associated nutrient requirements to meet that potential. The results of those trials demonstrated that the PDR potential as well as protein and water content of the genotype tested was higher than for slower growing genotypes. PDR was determined to be of the order of 100 g/d from 9 to 25 kg (Van Lunen and Cole, 1995a) and 175 - 187 g/d from 25 to 90 kg live weight (Van Lunen and Cole, 1994).

Assuming that lean contains 25% protein, the $PDR_{max}$ values observed in these trials suggested that lean gain was of the order of 400 g/d for pigs from 9 to 25 kg live weight, increasing to a peak of almost 750 g/d at 80 kg live weight followed by a rapid decline (Figure 12.9). PDR growth curves indicated that for a brief time PDR peaked at 238 g/d at 70 kg live weight in boars. Since this peak was very brief, it appears to have had little impact on mean PDR or response to lysine/ DE ratio over the growth period of 25 to 90 kg. However, special feeding strategies during that brief period may have positive effects on PDR in boars.

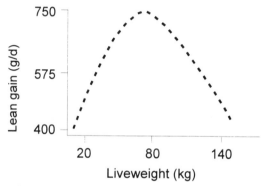

**Figure 12.9** Response of lean gain to live weight of pigs from 20 to 150 kg

The LDR of modern genotypes appears to be similar to slower growing genotypes. However, the higher PDR potential of modern genotypes results in lower lipid/protein ratios of gain and body composition. It should be noted, however, that the lipid/protein ratio of body composition increases with increasing weight in modern as well as slower growing genotypes.

P-2 measurements of the genotype tested were of the order of 4.5 mm at 20 kg live weight increasing to 14.5 mm at 150 kg live weight when fed a diet containing a lysine/DE ratio of 1.0 g/MJ (Figure 12.10).

From the curvilinear responses of DLWG and PDR to lysine/DE ratio it appears that dietary lysine/DE ratios above the optimum may be as detrimental to protein deposition as ratios below the optimum. As diets containing lysine/DE ratios above 1.0 are difficult and expensive to formulate on a commercial basis, this does not appear to be a practical concern.

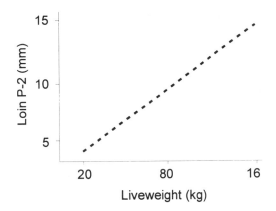

**Figure 12.10** P-2 measurements of pigs from 20 to 150 kg live weight

## FEEDING MODERN GENOTYPES

The daily lysine requirements for modern genotypes can be determined factorially from the PDR response to live weight. Table 12.7 shows the PDR and lysine requirements for maintenance and production of the pigs from 20 to 150 kg liveweight of Van Lunen and Cole (1995b). The maintenance requirement was taken as 36 mg/kg $W^{0.75}$ (Fuller *et al.*, 1987) while the lysine content of protein gain in the pig was taken to be 70 mg/g (Whittemore, 1993). True digestibility was taken to be 80% while absorption efficiency was assumed to be 70%. The daily lysine requirements listed in Table 12.7 are of the order of 14.25 g/d at 20 kg live weight to a maximum of 24.45 g/d at 80 kg live weight followed by a gradual decline. The values listed are similar to those of Close (1994a) and Friesen *et al.* (1994). It should be noted that maintenance requirements increased with increasing live weight, from 2.6% of the total requirement at 15 kg to 18.1% at 150 kg live weight.

Work previously reported on lysine/DE ratio requirements of slower growing pigs has given sex specific requirements. However, the results observed in the Nottingham trials suggested that sex, in modern genotypes, has only a minimal effect on PDR and nutrient requirements. In relation to live weight, boars and gilts exhibited similar PDR patterns in one study recently conducted at the University of Nottingham (Van Lunen and Cole, 1995b), while in another study boars, gilts and castrates responded similarly to dietary lysine/DE ratios in the weight range of 25 to 90 kg (Van Lunen and Cole, 1994).

Figure 12.11 shows the combined sex growth curve of body protein content to liveweight from 17 to 100 kg from the Nottingham work. It is apparent that protein content increased almost linearly during this growth phase. Optimum dietary lysine/DE ratio was determined at two weight ranges (10 - 25 and 25 - 90

**Table 12.7** LYSINE REQUIREMENTS AND OPTIMUM LYSINE/DE RATIO OF FAST GROWING PIGS FROM 15 TO 150 KG LIVE WEIGHT

| Live weight (kg) | PDR (g/d) | Maintenance[1] (g lysine/d) | Production[2] (g lysine/d) | Total (g lysine/d) | Optimum Lysine/DE (g/MJ) |
|---|---|---|---|---|---|
| 20 | 114 | 0.43 | 14.25 | 14.68 | 1.20 |
| 30 | 137 | 0.58 | 17.13 | 17.70 | 1.15 |
| 40 | 157 | 0.72 | 19.63 | 20.34 | 1.10 |
| 50 | 172 | 0.85 | 21.50 | 22.35 | 1.05 |
| 60 | 182 | 0.97 | 22.75 | 23.72 | 1.00 |
| 70 | 186 | 1.09 | 23.25 | 24.34 | 0.95 |
| 80 | 186 | 1.20 | 23.25 | 24.45 | 0.90 |
| 90 | 182 | 1.31 | 22.75 | 24.06 | 0.85 |
| 100 | 174 | 1.42 | 21.75 | 23.17 | 0.80 |
| 110 | 164 | 1.53 | 20.50 | 22.03 | |
| 120 | 152 | 1.63 | 19.00 | 20.63 | |
| 130 | 140 | 1.73 | 17.50 | 19.23 | |
| 140 | 127 | 1.83 | 15.88 | 17.71 | |
| 150 | 114 | 1.93 | 14.25 | 16.18 | |

[1] 36 mg/kg $W^{0.75}$, true digestibility 80%
[2] 70 mg/g lysine in PDR, true digestibility of 80%, absorption efficiency 70%.

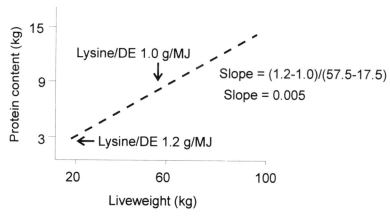

**Figure 12.11** Calculation of slope of optimum lysine/DE ratio to PDR

kg) and as the protein growth of the pigs was linear from 17 to 100 kg live weight, it appears that a linear equation can be used to estimate the lysine/DE requirements of fast growing pigs at specific weights within that range.  At a mean live weight of 17.5 kg (mean of 10 and 25 kg) the optimum lysine/DE ratio was determined as 1.2 g/MJ, while at a mean live weight of 57.5 kg (mean of 25 and 90 kg) the optimum ratio was found to be of the order of 1.0 g/MJ. The slope of the line between those two points was determined to be 0.005. Thus, the equation to predict the optimum lysine/DE ratio for pigs at a specific weight between 17 and 100 kg live weight would be:

$$Lysine/DE = (1.2 - 0.005(W\text{-}17.5)) \qquad (6)$$

This equation generated estimates of optimum dietary lysine/DE requirements of all sexes of fast growing genotypes (Table 12.8) from 20 to 100 kg liveweight. Values were not generated for liveweights above 100 kg as protein growth was not linear beyond that point.

**Table 12.8**   OPTIMUM LYSINE/DE (G/MJ) RATIOS IN THE DIETS OF GROWING PIGS

| Liveweight maximum PDR (g/d) | Sex | Unimproved 100 | Average 125 | High 150 | Hybrid 175 |
|---|---|---|---|---|---|
| Up to 25 kg | Castrate | 0.78 | 0.85 | 0.88 | 1.20 |
|  | Gilt | 0.80 | 0.85 | 0.90 | 1.20 |
|  | Boar | 0.83 | 0.88 | 0.93 | 1.20 |
| 25 to 55 kg | Castrate | 0.73 | 0.78 | 0.83 | 1.10 |
|  | Gilt | 0.75 | 0.80 | 0.85 | 1.10 |
|  | Boar | 0.78 | 0.83 | 0.88 | 1.10 |
| 55 to 90 kg restricted fed | Castrate | 0.55 | 0.55 |  |  |
|  | Gilt | 0.65 | 0.65 |  |  |
|  | Boar | 0.70 | 0.70 |  |  |
| 55 to 90 kg fed ad libitum | Castrate |  | 0.58 | 0.63 | 0.95 |
|  | Gilt |  | 0.60 | 0.65 | 0.95 |
|  | Boar |  | 0.63 | 0.68 | 0.95 |

After Cole (1992)

Although castrates have a greater tendency towards lipid deposition than boars with gilts being intermediate, all sexes of this genotype appeared to have lower target body lipid levels than those of slower growing genotypes. Because the protein content of live weight gain remained relatively constant for most of the growth period, DLWG and PDR patterns were similar and the equation:

$$PDR = 0.175DLWG \qquad (7)$$

could be used to estimate PDR from DLWG over the 25 to 90 kg live weight period assuming there were no nutritional limitations to PDR. This equation indicates that 17.5% of live-weight gain from 25 to 90 kg is in the form of protein in highly selected genotypes.

Equation 4 was not accurate at specific weights as PDR patterns changed with increasing age and weight. However as a general estimation for the growth period of 20 to 100 kg the equation is useful. It should also be noted that LDR did not reach a maximum until just prior to the slaughter weights of 90 to 100 kg.

The relationship of lysine/DE ratio to PDR of the pigs of the Nottingham work (Van Lunen and Cole, 1994) from 25 to 90 kg live weight is given in Figure 12.12. It demonstrates the effects of sub-optimal lysine/DE ratios on PDR of modern genotypes. For example, feeding a lysine/DE ratio of 0.8 g/MJ would result in a 10 g/d reduction in PDR while a similar reduction caused by oversupply of protein would occur.

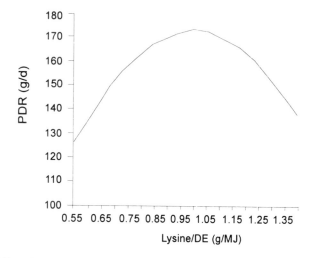

**Figure 12.12** Effect of dietary lysine/DE ratio on PDR of fast growing pigs from 25 to 90 kg liveweight

With the knowledge generated from work with modern genotypes, lysine/DE ratio requirements based on PDR potential can be estimated. Table 12.8 gives the recommended lysine/DE ratios of pigs of various PDR potentials. As discussed earlier, it appears that sex specific lysine/DE ratio recommendations are not required for modern fast growing pigs fed *ad libitum*. It is also clear that the higher PDR potential of the fast growing genotypes tested resulted in higher lysine/ DE ratio requirements than those reported by Cole (1992) for another high potential genotype.

Table 12.9 gives a summary and comparison of carcass lean gain, feed intake and nutrient requirements of the pigs of the Nottingham studies with values presented by Close (1994b) for a similar modern genotype. To allow for direct comparison to the Close (1994b) values, the total body lean gain values of the Nottingham work were multiplied by an assumed killing out percentage of 80%. In both cases lean was estimated from the protein content of the body or carcass. Feed intake values were similar for both studies suggesting that the Chadd, Cole and Walters (1993) equation: $DE\ (MJ/d) = 5.0W^{0.44}$ and the Close (1994b) equation: $DE = 1.5W^{0.57}$ describe the lower appetite levels of fast growing pigs. Daily lysine requirements were similar for both studies, although the Close (1994b) values are higher at higher live weights suggesting differences existed in the factorial approaches used to calculate the requirement. Lysine/DE ratio requirements were much higher at light weights for the Nottingham study. This may be due to the quadratic, rather than linear/plateau interpretation used.

**Table 12.9**  GROWTH PERFORMANCE, FEED INTAKE AND NUTRIENT REQUIREMENTS OF MODERN GENOTYPES

| Body weight (kg) | Carcass lean gain (g/d) | Feed intake (MJ/d) | Lysine requirement (g/d) | Lysine/DE (g/MJ) |
|---|---|---|---|---|
| 20 | 313 (293)[1] | 18.7 (15.5) | 14.7 (14.0) | 1.20 (0.90) |
| 40 | 384 (360) | 25.3 (24.5) | 20.3 (20.0) | 1.10 (0.82) |
| 60 | 432 (405) | 30.3 (30.0) | 23.7 (24.4) | 1.00 (0.81) |
| 80 | 456 (428) | 34.4 (34.0) | 24.5 (27.0) | 0.90 (0.79) |
| 100 | 480 (450) | 37.9 (36.5) | 23.2 (28.5) | 0.80 (0.78) |

[1] Values in parentheses from Close (1994b)

## Conclusion

Research with highly selected genotypes has improved understanding of the requirements for ideal protein and energy in relation to lean growth or PDR potential for all genotypes. Pigs with high PDR potentials have higher lysine/DE requirements as compared with slower growing pigs. Coupled with the reduced appetite of these new genotypes, it is imperative that energy levels be sufficient to minimize the PDR restriction which may be imposed by these lower appetite levels.

It is clear from work with slower growing pigs that lean growth potential and lysine/DE requirements are sex specific. Although highly selected pigs may still exhibit some sex difference in terms of lean growth potential, the differences appear to be of lesser magnitude, and as a result, lysine/DE ratio requirements to maximize that potential appear to be similar for all sexes.

# References

Agricultural Research Council (ARC) (1981) *The nutrient requirements of pigs.* Commonwealth Agricultural Bureaux, Slough.

Baker, D.H. and Chung, T.K. (1992) Ideal protein for swine and poultry. *BioKyowa Technical Review-4*, Chesterfield: NutriQuest, Inc.

Baker, D.H., Hahn, J.D., Chung, T.K. and Han, Y. (1993) Nutrition and growth: the concept and application of an ideal protein for swine growth. In: *Growth of the Pig*, Ed. G.R. Hollis. CAB International, Wallingford.

Batterham, E.S. (1974) The frequency of feeding on the utilization of free lysine by growing pigs. *British Journal of Nutrition* 39: 265-270.

Batterham, E.S. (1994) Protein and energy relationships for growing pigs. In: *Principles of Pig Science*, Eds. D.J.A. Cole, J. Wiseman, M.A. Varley, Nottingham University Press, Nottingham. pp 107–121.

Batterham, E.S. and O'Neill, G.H. (1978) Effect of frequency of feeding on response by growing pigs to supplements of free lysine. *British Journal of Nutrition* 64: 81-94.

Batterham, E.S., Anderson, L.M., Baigent, D.R. and White, E. (1990) Utilization of ileal digestible amino acids by growing pigs: effect of dietary lysine concentration on efficiency of lysine retention. *British Journal of Nutrition* 31: 237-348.

Batterham, E.S., Giles, L.R. and Dettmann, E.B. (1985) Amino acid and energy interactions in growing pigs. 1. Effect of food intake, sex and liveweight on the responses of growing pigs to lysine concentration. *Animal Production* 40:331-343.

Campbell, R.G. (1988) Nutritional constraints to lean tissue accretion in farm animals. *Nutrition Research Reviews* 1: 233-253.

Campbell, R.G. (1990) The effects of protein deposition capacity on the growing pig requirements for dietary nutrients. *Proceedings of Arkansas Nutrition Conference* Sept.12-14 pp. 117-124.

Campbell, R.G., Johnson, R.J., King, R.H. and Taverner, M.R. (1990) Effects of gender and genotype on the response of growing pigs to exogenous administration of porcine growth hormone. *Journal of Animal Sciience* 68: 2674-2681.

Campbell, R.G. and Taverner, M.R. (1985) Effect of strain and sex on protein and energy metabolism in growing pigs. In: *Energy Metabolism of Farm Animals European Association for Animal Production*. Eds. R.W. Moe H.F. Tyrell and P.J. Reynolds. Rowman and Littlefields USA pp. 78-81.

Campbell, R.G., Taverner, M.R. and Curic, D.M. (1984) Effects of feeding level and dietary protein content on the body composition and rate of protein deposition in pigs from 45 to 90 kg. *Animal Production* 38: 233-240.

Campbell, R.G., Taverner, M.R. and Curic, D.M. (1985) The influence of feeding level on the protein requirement of pigs between 20-45 kg. liveweight. *Animal Production* 40: 489-496.

Campbell, R.G., M.Taverner, M.R. and Curic, D.M. (1986) Effects of sex and energy intake between 48-90 kg liveweight on protein deposition in growing pigs. *Animal Production* 40: 497-503.

Campbell, R.G., Taverner, M.R., and Curic, D.M. (1988) The effects of sex and live weight on the growing pigs' response to dietary protein. *Animal Production* 46: 123-130.

Campbell, R.G., Taverner, M.R. and Mullaney, P.D. (1975) The effects of dietary concentrations of digestible energy on the performance and carcass characteristics of early weaned pigs. Animal Production 21: 285-294.

Carr, J.R., Boorman, K.N. and Cole, D.J.A. (1977) Nitrogen retention in the pig. *British Journal of Nutrition* 37: 143-155.

Chadd, S.A., Cole, D.J.A. and Walters, J.R. (1993) The food intake, performance and carcass characteristics of two pig genotypes grown to 120 kg liveweight. *Animal Production* 57: 473-481.

Chiba, L.I., Lewis, A.J., and Peo, E.R. (1991) Amino acid and energy interrelationships in pigs weighing 20-50 kg.:1. Rate and efficiency of weight gain. *Journal of Animal Sci*ence 69: 694-707.

Close, W.H. (1994a) Feeding new genotypes: establishing amino acid /energy requirements. In: *Principles of Pig Science.* Eds: D.J.A. Cole, J. Wiseman and M.A. Varley. Nottingham University Press, Nottingham. pp 123–140.

Close, W.H. (1994b) Nutitional needs and responses of the growing pig. Meeting the genetic potential for growth and lean tissue gain. *JSR Healthbred 5th Annual Technical Conference.*

Cole, D.J.A. (1978) Amino acid nutrition of the pig. In: *Recent Advances in Animal Nutrition.* Eds. W. Haresign and D. Lewis. Butterworths, London.

Cole, D.J.A. (1992) Interaction between energy and amino acid balance. *International Feed Production Conference*, Piacenza, Feb 25-26, 1992

Cole, D.J.A. and Chadd, S.A. (1989) Voluntary food intake of growing pigs. *Occasional Publication No. 13* British Society of Animal Production.

Cole, D.J.A., Duckworth, J.E. and Holmes, W. (1967) Factors affecting voluntary feed intake in pigs. II. The effect of two levels of crude fibre in the diet on the intake and performance of fattening pigs. *Animal Production* 9: 149-154.

Cooke, R., Lodge, G.A. and Lewis, D. (1972) Influence of energy and protein concentration in the diet on the performance of growing pigs. 1. Response to protein intake on a high energy diet. *Animal Production* 14: 35-44.

Cresswell, D.C., Wallace, H.D., Combs, G.E., Palmer, A.Z. and West, R.L. (1975) Lysine and tryptophan in diets for boars and barrows. *Journal of Animal Science* 40:167 Abstr.

Dunkin, A.C., Black, J.L. and James, K.J. (1986) Nitrogen balance in relation to energy intake in entire male pigs weighing 75 kg. *British Journal of Nutrition* 55:201-207.

Dunkin, A.C. and Black, J.L. (1985) The relationship between energy intake and nitrogen balance in the growing pig. In: *Energy Metabolism of Farm Animals*. eds. R.W. Moe, H.F.Tyrll and P.J.Reynolds. Rowman and Littlefield, U.S.A.

Ellis, M.,Smith, W.C., Henderson, R., Whittemore, C.T. and Laird, R. (1983) Comparative performance and body composition of control and selection line Large White pigs. 2. Feeding to appetite for a fixed time. *Animal Production* 36: 407-413.

English, P.R., Fowler, V.R., Baxter, S. and Smith,W. (1988) *The Growing-Finishing pig. Improving efficiency*. Farming Press, Ipswitch.

Ferguson, N.S. and Gous, R.M. (1993) Evaluation of pig genotypes. 2. Testing experimental procedure. *Animal Production* 56: 245-249.

Fowler, V.R. (1978) Biological models of quantifying growth and efficiency. In: *Patterns of Growth and Development in Cattle*. Eds. H. de Boer and J. Martin. Martinus Nighoff Publishers, The Hague.

Friesen, K.G., Nelssen, J.L., Unruh, J.A., Goodband, R.D., and Tokach, M.D. (1994) Effects of the interrelationship between genotype, sex, and dietary lysine on growth performance and carcass composition in finishing pigs fed to either 104 or 127 kilograms. *Journal of Animal Science* 72: 946-954.

Fuller, M.F. and Wang, T.C. (1987) Amino acid requirements of the growing pig. In: *Manipulating Pig Production*. Ed. J.L. Barnett. Australian Pig Science Assoc., Werribee, Australia.

Fuller, M.F., Livingston, R.M. Baird, B.A., and Atkinson, T. (1979) The optimal amino acid supplementation of barley for growing pigs. 1. Response of nitrogen metabolism to progressive supplementation. *British Journal of Nutrition* 41: 321-331.

Fuller, M.F., McWilliam, R. and Wang, T.C. (1987) The amino acid requirements of pigs for maintenance and for growth. *Animal Production* 44: 476-488.

Fuller, M.F., Wood, J., Brewer, A.C., Pennie, K. and McWilliam, R. (1986) The response of growing pigs to dietary lysine, as free lysine hydrochloride or in soya-bean meal, and the influence of food intake. *Animal Production* 43: 447-484.

Giles, L.R., Batterham, E.S., and Dettman, E.B. (1986) Amino acid and energy interactions in growing pigs: 2. Effects of food intake, sex and liveweight on responses to lysine concentration in barley based diets. *Animal Production* 42: 133-144.

Henderson, R., Whittemore, C.T., Ellis, M.,Smith, W.C., Laird, R. and Phillips, P. (1983) Comparative performance and body composition of control and selection line Large White pigs. 1. On a generous fixed feeding scale for a fixed time. *Animal Production* 36: 399-405.

Holmes, C.W., Carr, J.R. and Pearson, G. (1980) Some aspects of the energy and nitrogen metabolism of boars, gilts and barrows given diets containing different concentrations of protein. *Animal Production* 31: 279-289.

Kanis, E. (1988) Effect of average daily feed intake on production performance in growing pigs. *Animal Production* 46: 111-122.

Kanis, E., Nieuhof, G.J., de Greef, K.H., Van Der Hel, W.,Verstegen, M.W.A., Huisman, J. and Van Der Wal, P. (1990) Effects of recombinant porcine somatotropin on growth and carcass quality in growing pigs: interactions with genotype, gender and slaughter weight. *Journal of Animal Sci*ence 68: 1193-1200.

Kemm, E.H., Siebrits, F.K. and Barnes, P.M. (1990) A note on the effect of dietary protein concentration, sex, type and liveweight on whole-body amino acid composition of the growing pig. *Animal Production* 51: 631-634.

Kielanowski, J. (1969) Energy and protein metabolism in growing pigs. *Renta Cubano Cienc Agriculture* 3: 207:216.

Lindsay, D.B. (1983) Growth and Fattening. In: *Nutritional Physiology of Farm Animals*. Eds. J.A.F. Rook and P.C. Thomas. Longman Group Ltd. Harlow.

Nam, D.S. and Aherne, F.X. (1994) The effects of lysine/energy ratio on the performance of weanling pigs. *Journal of Animal Science* 72: 1247-1256.

Newell, J.A. and Bowland, J.P. (1972) Performance, carcass composition and fat composition of boars, gilts and barrows fed two levels of protein. *Canadian Journal of Animal Science* 52: 543-551.

Prescott, J.H.D. and Lamming, G.E. (1967) The influence of castration on the growth of male pigs in relation to high levels of dietary protein. *Animal Production* 9: 535-545.

Rao, D.S. and McCracken, K.J. (1990) Effects of protein intake on energy and nitrogen balance and chemical composition of gain in growing boars of high genetic potential. *Animal Production* 51: 389-397.

Rao, D.S., and McCracken, K.J. (1991) Effects of energy intake on protein and energy metabolism of boars of high genetic potential for lean growth. *Animal Production* 52: 499-507.

Siebrits, F.K. and Kemm, E.H. (1982) Body composition and energetic efficiency of lean and obese pigs. In: *Energy Metabolism of Farm Animals*. European

Association for Animal Production Publ. no. 29 pp. 237-240. Eds. A. Ehern and F. Sundsol. Infomasjonstenknikk, Norway.

Stranks, M.H., Cooke, B.C., Fairbairn, C.B., Fowler, N.G., Kirby, P.S., McCracken, K.J., Morgan, C.A (1988) Nutrient allowances for growing pigs. *Research and Development in Agriculture.* 5: 71-88.

Tanksley, T.D. and Knabe, D.A. (1984) Ileal digestibilities of amino acids in pig feed and use in formulating diets. In: *Recent Advances in Animal Nutrition.* Eds. W. Haresign and D.J.A. Cole. Butterworths, London.

Taverner, M.R., Campbell, R.G. and King, R.H. (1977) The relative protein and energy requirements of boars, gilts and barrows. *Australian Journal of Experimental Agriculture and Animal Husbandry.* 1: 134-145.

Van Lunen, T.A. and Cole, D.J.A. (1994) Effect of lysine/digestible energy ratio on growth performance and nitrogen deposition of hybrid boars, gilts and castrates. *Animal Production.* 58: 435 (Abstr.)

Van Lunen, T.A. and Cole, D.J.A. (1995a) Effect of lysine/DE energy ratio and energy density on growth performance of highly selected pigs from 9 to 25 kg liveweight. *Journal of Animal Science.* 73 Supp. 1: 181

Van Lunen, T.A. and Cole, D.J.A. (1995b) Growth and nitrogen deposition of hybrid pigs from 10 to 150 kg liveweight. *Journal of Animal Science.* 73 Supp. 1: 137

Wang, T.C. and Fuller, M.F. (1987) An optimal dietary amino acid pattern for growing pigs. *Animal Production* 44: 476 Abst.

Wang, T.C., and Fuller, M.F. (1990) The effect of plane of nutrition on the optimum dietary amino acid pattern for growing pigs. *Animal Production* 50: 155-164.

Whittemore, C.T. (1986) An approach to pig growth modeling. *Journal of Animal Science.* 63: 615-621.

Whittemore, C.T. (1993) *The Science and Practice of Pig Production.* Longman Scientific and Technical, Harlow.

Whittemore, C.T. and Fawcett, R.H. (1976) Theoretical aspects of a flexible model to simulate protein and lipid growth in pigs. *Animal Production* 22:87-96.

Whittemore, C.T., Tullis, J.B. and Emmans G.C. (1988) Protein growth in pigs. *Animal Production* 46: 437-445.

Williams, W.D., Cromwell, G.L., Stahly, T.S. and Overfield, J.R. (1984) The lysine requirement of the growing boar versus barrow. *Journal of Animal Science* 58: 657-665.

Wiseman, J. Jagger, S., Cole, D.J.A. and Haresign, W. (1991) The digestion and utilization of amino acids of heat treated fish meal by growing/finishing pigs. *Animal Production* 53: 215-225.

Yen, H.T., Cole, D.J.A., and Lewis, D. (1986a) Amino acid requirements of growing pigs 8. The response of pigs from 50-90 kg. liveweight to dietary ideal protein. *Animal Production* 43: 155-165.

Yen, H.T., Cole, D.J.A., and Lewis, D. (1986b) Amino acid requirements of growing pigs 9. The response of pigs from 25-55 kg liveweight to dietary ideal protein. *Animal Production* 43: 141-154.

# LIST OF PARTICIPANTS

The thirtieth Feed Manufacturers Conference was organised by the following committee:

Dr C. Brenninkmeijer (Hendrix' Voeders Bv)
Dr W.H. Close (Close Consultancy)
Dr J. Harland (Trident Feeds)
Dr S. Jagger (Dalgety Agriculture)
Dr D. Kitchen (Amalgamated Farmers Ltd.)
Dr J.R. Newbold (Bocm Pauls Ltd.)
Dr J. O'Grady (Iaws Group Plc)
Mr P. Poornan (Lys Mill Ltd.)
Mr J.R. Pickford
Mr P.G. Spencer (Bernard Matthews Plc)
Mr D.H. Thompson (Rightfeeds Ltd)
Mr J. Twigge (Trouw Nutrition)
Dr K.N. Boorman
Prof P.J. Buttery
Dr D.J.A. Cole
Dr J.M. Dawson
Dr P.C. Garnsworthy (Secretary)        } University of Nottingham
Dr W. Haresign (Chairman)
Prof G.E. Lamming
Dr A.M. Salter
Dr J. Wiseman

The conference was held at the University of Nottingham, Sutton Bonington Campus, 3rd-5th January 1996 and the committee would like to thank the authors for their valuable contributions. The following persons registered for the meeting:

| | |
|---|---|
| Adams, Dr C.A. | Kemin Europa NV, Industriezone Wolfstee, 2200 Herentals, Belgium |
| Albers, Mr N. | BASF Aktiengesellschaft, 76056 Ludwigshafen, Germany |
| Allder, Mr M.J. | Eurotec Nutrition Ltd, Elondale House, 5b Martins Lane, Witcham, Ely, Cambridge CB6 2LB |
| Allen, Mrs D. | 46A Mudford Road, Yeovil, Somerset BA21 4AB |
| Allen, Dr J.D. | Frank Wright Ltd, Blenheim House, Blenheim Road, Ashbourne, Derbys DE6 1HA |
| Allison, Mr R. | University of Nottingham, Sutton Bonington Campus, Loughborough, Leics LE12 5RD |
| Anderson, Mr K.R. | Duffield Nutrition, Saxlingham Thorpe Mills, Ipswich Road, Norwich NR15 1TY |
| Anderson, Mr W. | Sun Valley Poultry Ltd, Feed Mill, Tram Inn, Allensmore, Hereford HR2 |
| Antoniella, Dr M. | Neofarma, via Emilia km 18 n'1854, 47020 Longiano (FO), Italy |
| Aronen, Mr I. | Raisio Feed Ltd, FIN-21201 Raisio, Finland |

263

| | |
|---|---|
| Asbury, Mr J. | IFIF News, Stoke Road, Bishops Cleeve, Glos GL52 4RW |
| Atherton, Dr D. | Thomson & Joseph Ltd, 119 Plumstead Road, Norwich |
| Baker, Mr S.J.L. | Roche Products Ltd, Heanor Gate, Heanor, Derbys DE75 7SG |
| Ball, Mr A. | Roche Products Ltd, Heanor Gate, Heanor, Derbys DE75 7SG |
| Barnes, Miss E.J. | Frank Wright Ltd, Blenheim House, Blenheim Road, Ashbourne, Derbys DE6 1HA |
| Barrie, Mr M. | Elanco Animal Health, Chapel Hill, Basingstoke, Hampshire RG21 5SY |
| Bartram, Dr C. | Oldacres, Church Road, Bishops Cleve, Cheltenham, Glos |
| Bates, Mrs A. | Chapman Vitrition Ltd, Ryhall Road, Stamford, PE9 1TZ |
| Beard, Mr M. | University of Nottingham, Sutton Bonington Campus, Loughborough, Leics LE12 5RD |
| Beardsworth, Dr P.M. | Roche Products Ltd, Heanor Gate, Heanor, Derbys DE75 7SG |
| Beaumont, Mr D. | Laboratories Pancoma (UK) Ltd, Crompton Road Industrial Estate, Ilkeston, Derbyshire DE7 4BG |
| Beckerton, Dr A. | Roche Products Ltd, Heanor Gate, Heanor, Derbys DE75 7SG |
| Bedford, Dr M.R. | Finnfeeds International Ltd, High Street, Marlborough, Wilts |
| Beer, Mr J.H. | W & J Pye Ltd, Fleet Square, Lancaster LA1 1HA |
| Beer, Dr J.V. | 127 Church Road, Salisbury, Wilts SP1 1RB |
| Beesty, Mr C. | Lloyds Animal Health Ltd, Morton, Oswestry, Shrops |
| Bell, Miss J.F. | W & J Pye Ltd, Fleet Square, Lancaster LA1 1HA |
| Bentley, Mr R. | Central Laboratories, Wildmere Road, Wildmere Industrial Estate, Banbury, Oxon OX16 7XS |
| Bercovici, Dr D. | Eurolysine, 16 Rue Ballu, 65009 Paris, France |
| Berry, Mr M.H. | Berry Feed Ingredients Ltd, Chelmer Mills, New Street, Chelmsford, Essex CM1 1PN |
| Best, Mr P. | Pig International, 18 Chapel Street, Petersfield, Hants GU32 3DZ |
| Beudeker, Dr R.F. | Gist-Brocades FSD, Agr; Ingredients Group, PO Box 1, 2600 MA Delft, Netherlands |
| Blake, Dr J. | Highfield, Little London, Andover, Hants SP11 6JE. |
| Bole, Mr J. | David Patton Ltd, Milltown Mills, Monaghan, Rep. of Ireland |
| Boorman, Dr K.N. | University of Nottingham, Sutton Bonington Campus, Loughborough, Leics LE12 5RD |
| Borgida, Mr L.P. | COFNA, 25 Rue du Rempart, 37018 Tours Cedex, France |
| Bourne, Mr S | Alltech UK Ltd, 16/17 Abenbury Way, Wrexham Industrial Estate, Wrexham, Clwyd LL13 9UZ |
| Brackenbury, Miss J. | BOCM Pauls, Lindum Mill, Ashby Road, Shepshed, Loughborough, Leics LE17 4HH |
| Brenninkmeyer, Dr C. | Hendrix Voeders BV, PO Box 1, 5830 MA Boxmeer, Holland |
| Brooking, Ms P. | International Additives, Old Gorsey Lane, Wallasey, L44 4AH |

| | |
|---|---|
| Brooks, Prof P.H. | University of Plymouth, Seale-Hayne Faculty, Newton Abbot, Devon TW12 6NQ |
| Brophy, Mr A. | Alltech Ireland Ltd, 28 Cookstown Industrial Estate, Tallaght, Dublin 24 |
| Brown, Mr G.J.P. | Roche Products Ltd, Heanorgate, Heanor, Derbys DE75 7SG |
| Brown, Mr J.M. | Britphos Ltd, Rawdon House, Yeadon, Leeds LS19 7BY |
| Brown, Mr M. | BOCM Pauls, Olympia Mills, Barlby Road, Selby, Yorks YO8 7AF |
| Bruce, Dr D.W. | Devenish Feed Supplements, 96 Duncrue Street, Belfast BT3 9AR |
| Burt, Dr A.W.A. | Burt Research Ltd, 23 Stow Road, Kimbolton, Huntingdon, Cambs PE18 0HU |
| Burt, Dr R. | MAFF, Ergon House, c/o Nobel House, 17 Smith Square, London SW1P 3JR |
| Buttery, Prof P.J. | University of Nottingham, Sutton Bonington Campus, Loughborough, Leics LE12 5RD |
| Bywater, Dr R.J. | Pfizer Ltd, Europe House, Bancroft Road, Reigate, Surrey RH2 7RP |
| Campbell, Sir C. | Vice Chancellor, University of Nottingham, University Park, Nottingham |
| Carroll, Mr I. | Technavet Ltd, Park Street, Congleton, CW12 1ED |
| Carter, Mr T.J. | Anitox Ltd, 80 Main Road, Earls Barton, Northants NN6 0HJ |
| Caygill, Dr J.C. | M A F F, Agric. & Food Tech. Div, Nobel House, 17 Smith Square, London SW1P 3JR |
| Charles, Dr C.R. | ADAS Gleadthorpe, Meden Vale, Mansfield, Notts NG20 9PF |
| Charlton, Mr P. | Alltech UK Ltd, 16/17 Abenbury Way, Wrexham Industrial Estate, Wrexham, Clwyd LL13 9UZ |
| Chubb, Dr L.G. | 39 Station Road, Harston, Cambridge |
| Clay, Mr J. | Alltech UK Ltd, 16/17 Abenbury Way, Wrexham Industrial Estate, Wrexham, Clwyd LL13 9UZ |
| Close, Dr W. | Close Consultancy, 129 Barkham Road, Wokingham, Berkshire RG41 2RS |
| Cole, Dr D.J.A. | Nottingham Nutrition International, 14 Potters Lane, East Leake, Loughborough, Leceicester, LE12 6NQ |
| Cole, Mr J. | International Additives, Old Gorsey Lane, Wallasey, L44 4AH |
| Colenso, Mr J. | Trouw Nutrition, Wincham, Northwich, Cheshire |
| Collyer, Mr M. | Kemin (UK) Ltd, Becor House, Green Lane, Lincoln LN6 7DL |
| Connolly, Mr J.G. | Red Mills, William Connolly & Sons Ltd, Gorsebridge, Co. Kilkenny, Ireland |
| Cooke, Dr B. | Dalgety Agriculture Ltd, 180, Aztec West, Almondsury, Bristol BS12 4TH |
| Cooper, Dr A. | University of Plymouth, Seale-Hayne Faculty, Newton Abbot TQ12 6NQ |
| Cottrill, Dr B. | ADAS, 'Woodthorne', Wolverhampton |
| Cowan, Dr D. | Novo Nordisk, 282 Chartridge Lane, Chesham, Bucks HP5 2SG |
| Cox, Mr N. | SCA Nutrition, Maple Mill, Dalton Airfield Industrial Estate, Dalton, W. Yorkshire |

| | |
|---|---|
| Creasey, Mrs A. | BASF plc, PO Box 4, Earl Road, Cheadle Hulme SK8 6QG |
| Davies, Dr J.L. | Roche Products Ltd, Heanor Gate, Heanor, Derbys DE75 7SG |
| Dawson, Dr J.M. | University of Nottingham, Sutton Bonington Campus, Loughborough, Leics LE12 5RD |
| Dawson, Mr W. | Britphos Ltd, Rawdon House, Green Lane, Yeadon, Leeds LS19 7BY |
| De Smet, Mr P. | Alltech Netherlands BV, |
| Deaville, Mr E. | ADAS, Feed Evaluation & Nutrition, Dairy Research Centre, Stratford upon Avon |
| Deaville, Mr S. | Rumenco, Stretton House, Derby Road, Burton on Trent |
| Dickins, Mr A.C. | Unitrition International Ltd, Olympia Mills, Barlby Road, Selby, N Yorks YO8 7AF |
| Dixon, Mr D.H. | Brown & Gillmer Ltd, PO Box 3154, The Lodge, Florence House, 199 Strand Road, Dublin 4, Ireland |
| Doran, Mr B. | Trouw Nutrition, Wincham, Northwich, Cheshire CW9 6DF |
| Drakley, Ms C. | University of Nottingham, Sutton Bonington Campus, Loughborough, Leics LE12 5RD |
| Edwards, Miss S. | Trouw Nutrition, Wincham, Northwich |
| Ewing, Mrs A. | Oldacres, Church Road, Bishops Cleeve, Cheltenham, Glos |
| Ewing, Dr W. | Cargill plc, Camp Road, Swinderby, Lincs |
| Farley, Mr R. | Trouw Nutrition UK, Wincham, Northwich |
| Fawthrop, Mr G. | Eurotec Nutrition Ltd, Glendale House, 5b Martins Lane, Witchem, Ely, Cambridge CB6 2LB |
| Filmer, Mr D. | David Filmer Ltd, Wascelyn, Brentknoll, Somerset TA9 4DT |
| Fitt, Dr T. | Roche Products Ltd, Heanor Gate, Heanor, Derbyshire DE4 5BJ |
| Fledderus, Ir J. | Cooperatie ABC ua, Kwinkweerd 12, NL-7241 CW Lochem, Netherlands |
| Fletcher, Mr C.J. | Aynsome Laboratories Ltd, Eccleston Grange, Prescot Road, St Helens, Merseyside WA10 3BQ |
| Flint, Prof A.P.F. | University of Nottingham, Sutton Bonington Campus, Loughborough Leics LE12 5RD |
| Flower, Mr A. | M A F F, Ergon House, Smith Square, London |
| Fordyce, Mr J. | WMF Ltd, Bradford Road, Melksham, Wilts SN12 8LQ |
| Foulds, Mr S. | Park Tonks Ltd, 48 North Road, Great Abington, Cambridge CB1 6AS |
| Fullarton, Mr P.J. | Forum Products Ltd, 41-51 Brighton Road, Redhill, Surrey RH1 6YS |
| Gardner, Miss N. | University of Reading, Reading, Berks RG6 2AH |
| Garland, Mr P. | BOCM Pauls Ltd, 47 Key Street, Ipswich, IP4 1BX |
| Garnsworthy, Dr P.C. | University of Nottingham, Sutton Bonington Campus, Loughborough, Leics LE12 5RD |
| Geddes, Mr N. | Nutec Ltd, Eastern Avenue, Lichfield, Staffs |

| | |
|---|---|
| Gibson, Mr J.E. | Parnutt Foods Ltd, Hadley Road, Woodbridge Industrial Estate, Sleaford, Lincs NG34 7EG |
| Gibson, Mr W.W.C. | Greenbank, Clarencefield, Dumfries. DG1 4NF |
| Gilbert, Mr R. | Asbury Publications, Stoke Road, Bishops Cleeve, Glos GL52 4RW |
| Gillespie, Miss F. | United Molasses, Derby Road, Stretton, Burton-on-Trent, Staffs DE13 0DW |
| Glennon, Mrs A. | N O A H, 3 Crossfield Chembers, Gladbeck Way |
| Goldsbrough, Mr T. | Daylay Foods Ltd, The Mill, Seamer, Stokesley, Middlesbrough, Cleveland TS9 5NQ |
| Gorless, Mr J. | Trouw Nutrition, 36 Ship Street, Belfast BT15 1JL |
| Gould, Mrs M. | Volac International Ltd, Orwell, Royston, Herts |
| Grace, Mr J. | Elanco Animal Health, Dextra Court, Chapel Hill, Basingstoke, Hamps RG21 5SY |
| Graham, Mr M.R. | Intermol, 19 Sandhills Lane, Liverpool L5 9XE |
| Gray, Mr W. | Kemira Kemi UK Ltd, Orm House, Hookstone Park, Harrogate, Yorks |
| Griffiths, Mr D. | Midland Shires Farmers Ltd, Defford Mill, Earls Croome, Worcester WR8 9DF |
| Hardwick, Ms J. | Grampian Pharmaceuticals, Marathon Place, Moss Side Industrial Estate, Leyland, Lancs |
| Haresign, Dr W. | University of Nottingham, Sutton Bonington Campus, Loughborough, Leics LE12 5RD |
| Harker, Dr A. | Finnfeeds International Ltd, High Street, Marlborough, Wilts |
| Harland, Dr J. | Dalgety Agriculture Ltd, 180 Aztec West, Almondsbury, Bristol BS12 4TH |
| Harrison, Mr M. | Farmlab, Whetstone Magna, Lutterworth Road, Leicester |
| Haythornthwaite, Mr A. | Nu Wave, Wild Goose House, Goe Lane, Freckleton, Preston, Lancs PR4 1HX |
| Hazzledine, Mr M. | Dalgety Agriculture Ltd, 180 Aztec West, Almondsbury, Bristol BS12 4TH |
| Higginbotham, Dr J.D. | United Molasses, Derby Road, Stretton, Burton-on- Trent, Staffs DE13 0DW |
| Hobson, Mr P. | Sciantec Analytical Services, Main Site, Dalton, Thirsk, N Yorks |
| Hockey, Mr R. | Pfizer Ltd, Ramsgate Road, Sandwich Kent CT13 8NJ |
| Hohn, Mr H.J. | Elanco Animal Health, Dextra Court, Chapel Hill, Basingstoke, Hamps RG21 5SY |
| Holder, Mr P. | Intermol, Shell Road, Royal Edward Dock, Avonmouth, Bristol BS11 9BW |
| Holmes, Mr G.F. | S & E Johnson Ltd, Old Road Mills, Darley Dale, Matlock, Derbys DE4 2ES |
| Horler, Ms R. | Sun Valley Poultry Ltd, Feed Mill, Tram Inn, Allensmore, Hereford HR2 9AW |

| | |
|---|---|
| Horner, Mr B. | Roche Products Ltd, Heanor Gate, Heanor, Derbys DE75 7SG |
| Houseman, Mr R. | Candus Ltd, 23B Metcalfe House, Kirkgate, Ripon |
| Howie, Mr A. | Nutrition Trading (Int.)Ltd, Orchard House, Manor Drive, Morton Bagot, Studley, Warks |
| Hughes, Mr D.P. | NWF Agriculture Ltd, Wardle, Nantwich, CW5 6AQ |
| Hughes, Prof P.E. | University of Melbourne, Dept. Agriculture & Resource Management, Parkville, Vic. 3052, Australia |
| Ingham, Mr R.W. | Kemin (UK) Ltd, Becor House, Green Lane, Lincoln LN6 7DL |
| Jacklin, Mr D. | Keenan TMR Centre, NAC, Stoneleigh Park, Kenilworth, Warwickshire |
| Jackson, Mr J.C. | Nutec Ltd, Eastern Avenue, Lichfield, Staffs |
| Jagger, Dr S. | Dalgety Agriculture Ltd, 180 Aztec West, Almondsbury, Bristol NS12 4TH |
| Janes, Mr R. | Criddle Billington Feeds, Warrington Road, Glazebury, Cheshire |
| Jardine, Mr G. | Guttridge Milling, 1 Mount Terrace, York YO2 4AR |
| Johnson, Miss S. | Kemira Kemi UK Ltd, Orm House, Hookstone Park, Harrogate, Yorks |
| Johnston, Mr K. | American Protein Corporation, 2 Silverwood Industrial Estate, Craigavon BT66 6EN |
| Jones, Dr E. | Dalgety Agriculture Ltd, 180 Aztec West, Almondsbury, bristol BS12 4HQ |
| Jones, Mr H. | Heygate & Sons Ltd, Bugbrooke Mills, Bugbrooke, Northampton NN7 3QH |
| Kahrs, Dr D. | Lohmann-LTE-GmbH, Postfach 446, D-27454 Cuxhaven |
| Keeling, Mrs S. | Nottingham University Press, Manor Farm, Thrumpton, Notts |
| Kemp, Mr G. | Orffa Nederland Feed BV, Burgstraat 12, 4283 GG Giessen, The Netherlands |
| Kenyon, Mr P.W. | Harbro Farm Sales Ltd, Markethill, Turriff, Aberdeenshire AB53 7PA |
| Keys, Mr J. | 32 Holbrook Road, Stratford-on-Avon, Warwickshire |
| Kitchen, Dr D. | Amalgamatd Farmers Ltd, Kinross, Newhall Lan, Preston PR1 5JX |
| Laird, Mr M.I. | FLD Chemical Distributors Ltd, Calbrook House, Eastwick Road, Bookham, Surrey KT23 4DT |
| Lake, Mr P. | J Bibby Agriculture Ltd, ABN House, PO Box 250, Oundle Road, Woodston,Peterborough PE2 9GF |
| Lamming, Prof E. | University of Nottingham, Sutton Bonington Campus, Loughborough, Leics LE12 5RD |
| Law, Mr J. | Sheldon Jones Agric, Portbury Mill, Royal Portbury Dock, Bristol |
| Lawrence, Dr K. | Elanco Animal Health, Dextra Court, Chapel Hill, Basingstoke, Hampshire RG21 5SY |
| Leaver, Prof J.D. | Wye College, Ashford, Kent TN25 5AH |
| Lee, Dr P. | ADAS Rosemaund, Preston Wynne, Hereford HR1 3PG |

| | |
|---|---|
| Long, Mr M.C.A. | The Flag Partnership, Chaucer House, Chaucer Road, Sudbury, Suffolk CO10 6LN |
| Lowe, Mr J. | Gilbertson & Page, PO Box 321, Welwyn Garden City, Herts |
| Lowe, Dr R.A. | Frank Wright Ltd, Blenheim House, Blenheim Road, Ashbourne, Derbys DE6 1HA |
| Lucey, Mr P. | Dairygold Quality Feeds, Lombardstown, Mallow, Co Cork |
| Lyons, Dr P. | Alltech Inc., Biotechnology Centre, 3031 Catniphill Pike, Nicholasville, KY 40356, USA |
| MacDonald, Mr P. | Alltech UK Ltd, 16/17 Abenbury Way, Wrexham Industrial Estate, Wrexham, Clwyd LL13 9UZ |
| Mafo, Mr A. | Proctors (Bakewell) Ltd, Hi Peak Feeds Mill, 12 Ashbourne Road, Derby DE22 3AA |
| Main, Mr J. | Tithebarn Limited, PO Box 20, Weld Road, Southport |
| Malandra, Dr F. | Sildamin, Sostegno di Spessa, 27010 Pavia, Italy |
| Marchment, Dr S. | Dalgety Ltd, 180 Aztec West, Almondsbury, Bristol BS12 4TH |
| Marsden, Dr M. | J Bibby Agriculture Ltd, ABN House, PO Box 250, Peterborough |
| Marsden, Dr S. | Dalgety Agriculture Ltd, 180 Aztec West, Almondsbury, Bristol BS12 4TH |
| Marsh, Mr S.P. | Rumenco, Stretton House, Derby Road, Burton-on-Trent, Staffs DE13 0DW |
| Mayne, Dr S. | ARINI, Hillsborough, Co Down, BT26 6DR, N Ireland |
| McCord, Mrs F. | John Thomson & Sons Ltd, 35-39 York Road. Belfast BT15 3GW |
| McGilloway, Dr D. | ARINI, Large Park, Hillsborough, Co. Down BT26 6DR |
| McLean, Mr D.R. | W L Duffield & Sons Ltd, Saxlingham Thorpe Mills, Norwich, Norfolk |
| Mills, Mr C. | University of Nottingham, Sutton Bonington Campus, Loughborough, Leics LE12 5RD |
| Millward, Mr J. | The Royal Pharmaceutical Soc., of Great Britisin, 1 Lambeth High St., London SE1 7JN |
| Morgan, Mr J | Lloyds Animal Health Ltd, Morton, Oswestry, Shrops. |
| Mounsey, Mr A.D. | HGM Publications, Abney House, Baslow, Bakewell, Derbyshire DE45 1RZ |
| Mounsey, Mr H.E. | HGM Publications, Abney House, Baslow, Bakewell, Derbyshire DE45 1RZ |
| Mounsey, Mr S.P. | HGM Publications, Abney House, Baslow, Bakewell, Derbyshire DE45 1RZ |
| Murray, Mr F. | Dairy Crest Ingredients, Philpot House, Rayleigh, Essex |
| Murray, Dr I | Scottish Agricultural College, 581 King Street, Aberdeen AB9 1UD |
| Neelsen, Mr T. | Alltech Netherlands BV, Hollandsch Diep 63, 2904 EP Cappelle aan den IJ5 5EL, Netherlands |
| Nelson, Ms J. | UKASTA, 3 Whitehall Court, London SW1A 2EQ |

| | |
|---|---|
| Newbold, Dr J. | BOCM Pauls, 47 Key Street, Ipswich, IP4 1BX |
| Newcombe, Mrs J.O. | University of Nottingham, Sutton Bonington Campus, Loughborough, Leics LE12 5RD |
| Nixey, Dr C. | British United Turkeys Ltd, Hockenhull Hall, Tarvin, Cheshire CH3 8LE |
| Noblet, Dr J. | INRA, Station de Recerches Porcines, 35590 - St Gilles, France |
| O'Brien, Mr J. | Vet. Medicines Directorate, Woodham Lane, New Haw, Addlestone, Surrey KT15 3NB |
| O'Connell, Mr C. | Keenan TMR Centre, NAC, Stoneleigh Park, Kenilworth, Warwickshire |
| O'Grady, Dr J. | IAWS Group plc, 151 Thomas Street, Dublin 8, Ireland |
| Oram, Mr | Roche Products Ltd, Heanor Gate, Heanor, Derbys DE75 7SG |
| Overend, Dr M.A. | Nutec Ltd, Eastern Avenue, Lichfield, Staffs |
| Packington, Mr A.J. | Roche Products Ltd, Heanor Gate, Heanor, Derbys DE75 7SG |
| Papasolomontos, Dr S. | Dalgety Agriculture Ltd, 180 Aztec West, Almondsbury, Bristol BS12 4TH |
| Partridge, Dr G. | Finnfeeds International Ltd, High Street, Marlborough, Wilts |
| Pass, Mr R. | United Distillers, 33 Ellersly Road, Edinburgh EH12 6JW |
| Pattison, Dr H. | Roche Products Ltd, Heanor Gate, Heanor, Derbys DE75 7SG |
| Payne, Mr J.D. | Borregaard Lignotech, PO Box 162, N-1701 Sarpsborg, Norway |
| Pehrson, Dr B. | Swedish Univ. of Agric. Sci., Experimental Station, PO Box 237, S-53223 Skara, Sweden |
| Peisker, Dr M. | ADM BioProducts, ADM Olmuhlen GmbH, Auguste - Viktiria Str 16, D-65185 Wiesbaden, Germany |
| Percival, Dr D. | SAC, Grassland & Ruminant Sciences Dept., Auchincruive, Ayr, Scotland KA6 5HW |
| Perrott, Mr J.G. | Trident Feeds, PO Box 11, Oundle Road, Peterborough PE5 7AR |
| Petersen, Dr S. | University of Nottingham, Sutton Bonington Campus, Loughborough, Leics |
| Pettersson, Mr L. | Svenska Foder AB, Box 673, S-53116 Lidkoping, Sweden |
| Phillips, Mr G. | Silo Guard Europe, Greenway Farm, Charlton Kings, Cheltenham |
| Pickess, Mr K. | Elanco Animal Health, Dextra Court, Chapel Hill, Basingstoke, Hamts. RG21 5SY |
| Pickford, Mr B. | Bocking Hall, Bocking Church Street, Braintree, Essex CM7 5JY |
| Pike, Dr I.H. | IFOMA, 2 College Yard, Lower Dagnall Street, St Albans, Herts AL3 4PA |
| Plowman, Mr G.B. | G W Plowman & Son Ltd, Selby House, High Street, Spalding, Lincs |
| Poornan, Mr P. | Lys Mill Ltd, Watlington, Oxon OX9 5ES |
| Povey, Dr G.M. | J Bibby Agriculture, PO Box 250, Oundle Road, Peterborough |
| Powles, Dr J. | Harper Adams Agric. College, Newport, Shropshire TF10 8NB |

| | |
|---|---|
| Probert, Ms L. | University of Nottingham, Sutton Bonington Campus, Loughborough, Leics LE12 5RD |
| Putnam, Mr M. | Roche Products Ltd, Heanor Gate, Heanor, Derbys DE75 7SG |
| Rae, Dr R. | Premier Nutrition Products Ltd, The Levels, Rugeley, Staffs WS15 1RD |
| Raine, Dr H. | ABN, PO Box 250, Oundle Road, Woodston, Peterborough PE2 9QF |
| Raper, Mr G.J. | Laboratories Pancosma (UK) Ltd, Crompton Road Industrial Estate, Ilkeston, Derbyshire DE7 |
| Redshaw, Dr M. | Degussa Ltd, Winterton House, Winterton Way, Macclesfield, SK11 0LP |
| Reeve, Dr A. | ICI Nutrition, Alexander House, Crown Gate, Runcorn WA7 2UP |
| Reeve, Mr J. | RS Feed Blocks, Orleigh Mill, Bideford, Devon |
| Revett, Mr S.D. | Aspland & James Ltd, 118 Bridge Street, Chatteris, Cambs |
| Roberts, Mr J.C. | Harper Adams Agric. College, Newport, Shrops. TF10 8NB |
| Robertshaw, Miss K. | BOCM Pauls Ltd, 47 Key Street, Ipswich IP4 1BX |
| Robinson, Mr D.K. | Favor Parker Ltd, The Hall, Stoke Ferry, King's Lynn, Norfolk |
| Roet, Mr R. | Novus UK Ltd, Cotteswold House, 14 Gloucester Street, Cirencester, Glos GL7 2DG |
| Rosen, Dr G. | Consultant, 66 Bathgate Road, London SW19 5PH |
| Russell, Ms S. | Farmlab, Whetstone Magna, Lutterworth Rd, Leicester |
| Ryan, Mr M. | Marigot Ltd, Celtic Sea Minerals, Marina Commercial Park, Co Cork, Ireland |
| Salter, Dr A.M. | University of Nottingham, Sutton Bonnington Campus, Loughborough, Leics LE12 5RD |
| Schaper, Mr S. | CUB, Runderweg 6, 8219 PK Lelystad, The Netherlands |
| Schulze, Dr H. | Finnfeeds International Ltd, High Street, Marlborough, Wilts |
| Scott, Prof R.K. | University of Nottingham, Sutton Bonington Campus, Loughborough, Leics LE12 5RD |
| Shepherd, Mr A.L. | 89 Kingston Hill, Kingston upon Hull, Surrey KT2 7PZ |
| Shorrock, Dr C. | FSL Bells, Hartham, Corsham SN13 0QB |
| Short, Ms F. | University of Nottingham, Sutton Bonington Campus, Loughborough, Leics LE12 5RD |
| Shrimpton, Dr D.H. | International Milling, Turret House, 171 High Street, Rickmansworth WD3 1SN |
| Shurlock, Dr T.G.H. | Lucta SA, PO Box 112. 08080 Barcelona, Spain |
| Silvester, Miss L. | Dalgety Agriculture Ltd, 180 Aztec West, Almondsbury, Bristol BS12 4TH |
| Sinclair, Dr L.A. | Harper Adams Agric. College, Newport, Shropshire TF10 8NB |
| Sketcher, Mrs S. | Trouw Nutrition, Wincham, Northwich |
| Sloan, Dr B.K. | Rhone Poulenc Animal Nutrtion, 42 Avenue Aristae Briand, BP100 Antony, Cedex, France |

| | |
|---|---|
| Sloan, Mr J. | Roche Products Ltd, 25 Stockmans Way, Belfast BT9 7JK |
| Smith, Mr E.C. | Farmway Ltd, Cock Lane, Piercebridge, Co Durham |
| Spencer, Mr A.R. | Roche Products Ltd, Heanor Gate, Heanor, Derbys DE75 7SG |
| Spencer, Mr P.G. | Bernard Matthews plc, Gt Witchingham Hall, Norwich NR9 5QD |
| Spreeuwenberg, Mr W.M.M. | Cehave NV, Postbus 200, 5460 BC Veghel, Netherlands |
| Stahly, Prof T.S. | Iowa State University, Dept. Animal Science, Ames, Iowa 50011, USA |
| Stainsby, Mr A.K. | BATA Ltd, Norton Road, Malton, N Yorks YO17 0NU |
| Stebbens, Dr H.R. | Park Tonks Ltd, 48 North Road, Great Abington, Cambridge CB1 6AS |
| Stein, Mr L | Felleskjopet Rogaland Agder, PO Box 208, 4001 Stavanger, Norway |
| Sterten, Mr H. | Felleskjopet Forutvikling, N-7005 Trondheim, Norway |
| Storey, Mr J. | Volac International Ltd, Orwell, Royston, Herts |
| Sumner, Dr R. | Midland Shires Farmers Ltd, Defford Mill, Earls Croome, Worcester WR8 9DF |
| Sylvester, Mr D. | Roche Products Ltd, Heanor Gate, Heanor, Derbys DE75 7SG |
| Taylor, Dr A.J. | Roche Products Ltd, Heanor Gate, Heanor, Derbys DE75 7SG |
| Taylor, Dr S.J. | Volac International Ltd, Orwell, Royston, Herts |
| Thomas, Dr C | SAC, Grassland & Ruminant Science Dept., Auchincruive, Ayr KA6 5HW |
| Thomas, Mr G. | Crown Chicken Ltd, Kenninghall, Norfolk |
| Thompson, Mr D.D. | Rightfeeds Ltd, Castlegarde, Cappamore, Co Limerick, Ireland |
| Threlfall, Dr E.J. | PHLS, Lab of Enteric Pathogens, 61 Colindale Avenue, London NW9 5HT |
| Tibble, Mr S.J. | SCA Nutrition Ltd, Maplemill, Dalton Industrial Estate,Dalton, W Yorkshire |
| Trebble, Mr J.W. | Mole Valley Farmers, Station Road, South Molton, Devon EX36 3BH |
| Tuck, Mr K. | Alltech Ireland Ltd, 28 Cookstown Industrial Estate, Tallaght, Dublin 24, Ireland |
| Twigg, Mr J. | Trouw Nutrition, Wincham, Northwich, Cheshire |
| Van Der Aar, Mr P.J. | De Schothorst, Postbox 533, NL-8200 AM, Lelystad, Netherlands |
| Van Der Klis, Dr J.D. | DDLO, Runderweg 2, Postbus 160, 8200 Lelystad, The Netherlands |
| Van Der Ploeg, Mr H. | Stationsweg 4, 3603 EE, Maarssen, Netherlands |
| Van Krimpen, Mr M. | Cavo Latuco, PO Box 40262, 3504 AB Utrecht, Netherlands |
| Van Lunen, Dr T | Univ. of Prince Edward Island, 550 University Ave., Charlottetown, P E I, C1A 4P3, Canada |
| Vernon, Dr B. | BOCM Pauls Ltd, PO Box 339, 47 Key Street, Ipswich IP4 1BX |
| Vestervall, Dr F. | Lantmennen Feed, Box 30192, S-10425 Stockholm, Sweden |
| Vik, Mr K.R. | Stormollen AS, 5270 Vaksdal, Norway |

| | |
|---|---|
| Wales, Mr C. | Dalgety Agriculture Ltd, 180 Aztec West, Almondsbury, Bristol BS12 |
| Walton, Dr J. | University of Liverpool, Veterinary Field Station, Neston, South Wirral L64 7TE |
| Wareham, Dr C.N. | Grain Harvesters Ltd, The Old Colliery, Wingham, Canterbury, Kent |
| Waters, Dr C.J. | Trident Feeds, PO Box 11, Oundle Road, Peterborough PE5 7AR |
| Webster, Mrs M. | Format International, FormatHouse, Poole Road, Woking, Surrey GU21 1DY |
| Whyte, Dr A. | Rowett Research Services, Greenburn Road, Bucksburn, Aberdeen AB2 9SB |
| Wilkinson, Prof M. | De Montfort University, Riseholme Hall, Lincoln LN2 2LG |
| Wilkinson, Dr R.G. | Harper Adams Agric. College, Newport, Shropshire TF10 8NB |
| Williams, Mr C. | Trouw Nutrition, Wincham, Northwich, Cheshire |
| Williams, Mr D. | Intermol, Shell Road, Royal Edward Dock, Avonmouth, Bristol BS11 9BW |
| Williams, Dr D.R. | Anitox Ltd, 80 Main Road, Earls Barton, Northants NN6 0HJ |
| Williams, Mr J.S. | Roche Products Ltd, Heanor Gate, Heanor, Derbys DE75 7SG |
| Williams, Miss N. | Harbro Farm Sales, Turriff, Aberdeenshire AB53 7PA |
| Williams, Mr P.G. | AKZO Nobel Surface Chemistry, 23 Grosvenor Road, St Albans, Herts AL1 3AW |
| Williams, Mr W.I. | 40 Beech Avenue, Worcester WR3 8PY |
| Wilson, Dr B.J. | Cherry Valley Farms Ltd, Divisional Offices, North Kelsey Moor, Caistor, Lincs LN7 6HH |
| Wiseman, Dr J. | University of Nottingham, Sutton Bonington Campus, Loughborough, Leics LE12 5RD |
| Woolford, Dr M. | Alltech UK Ltd, 16/17 Abenbury Way, Wrexham Industrial Estate, Wrexham, Clwyd LL13 9UZ |
| Youdan, Dr J. | Nutrimix, Boundary Industrial Estate, Boundary Road, Lytham, Lancs FY8 5HU |
| Zwart, Mr J.E.M. | Tessenderlo Chemie Rotterdam, Maassluissedijk 103, 3133 KA Vlaardingen, The Netherlands |

# INDEX

# CUMULATIVE CONTENTS - VOLUMES 1–30

Numbers refer to volume.page, e.g. 21.117 means volume 21, page 117.

279

# 284 *Cumulative contents - volumes 1–30*

# CUMULATIVE INDEX - VOLUMES 1–30

Numbers refer to volume.page, e.g. 21.117 means volume 21, page 117. Papers are indexed by title only and according to main species and main subject.